Drying, Curing
and
Smoking Food

DRYING, CURING AND SMOKING FOOD

Marian Faux

Grosset & Dunlap
A Filmways Company
Publishers • New York

CONTENTS

5

Drying, Curing and Smoking Food

1

DRYING FOOD
AT HOME –
AN ANCIENT ART
YOU TOO CAN ENJOY

DRYING FOOD
AT HOME –
AN ANCIENT ART
YOU TOO CAN ENJOY

A generation of Americans who have grown up on such modern prepackaged foods as bouillon cubes, a confusing array of biscuits and crackers, and a snack food made of the dried beef known as "jerky" may be surprised to learn that the ancient home art of drying foods is making a comeback. And they may be even more surprised to learn that all the foods just mentioned are indeed dried foods.

Drying food is truly an art of the people. Unlike canning, there are few rules to guide the home drier, and each drying experience varies, depending upon the weather, the quality of food dried, and the technique used. Fortunately, though, you can easily and enjoyably acquire skills in drying food, primarily through your own personal experience. In a sense, drying food is akin to growing your own garden. There are procedures to learn and follow for success, but there is also room for creativity.

When Did Drying Foods Start?

Dried food was part of the lives of prehistoric peoples, who recognized, as do most home driers today, that herbs and spices were among the easier foods to dry. Poppy and caraway seeds have been found in Swiss prehistoric sites. Herbs were cultivated in early Assyrian and Babylonian gardens, and dried forms of these plants have been found in their ancient tombs.

Dates, figs, and grapes, among other fruits, were important in the diet of prehistoric people; they appear frequently in their art and in their tombs.

By the time of the Roman historian, Pliny the Elder (A.D. 23–79), apples were cultivated for their long-lasting qualities, and there were 36 varieties of apple pie. It is safe to assume that more than a few of the 36 recipes used dried apples. Along with apples, figs, pomegranates, and dates, pears were cultivated in Homeric times—and their value as a dried food staple was not overlooked.

On the other side of the world, several Indian cultures were based on dried foods such as corn, acorns (which were ground into an excellent flour), and berries. One South American Indian culture existed on *chuno*, a dried potato that was achieved by first exposing the potatoes to frost at night, then trampling them with one's feet, and finally sun-drying them for several days. Sometimes the process was varied slightly to make *tunta*, a white *chuno* that could be converted to flour.

The Process of Drying

The theory behind drying is simple. Drying is the process of removing enough water from certain foods to halt or slow down the growth of damaging organisms. The process is natural because no chemicals are added to the food during drying. Although some amounts of vitamins C and A are lost through drying and storage, dried food, pound for pound, has a greater concentration of nutrients, especially minerals, than other foods. In addition, the quicker the process of drying, the less chance that nutrients will be lost.

The amount of moisture left in a food depends upon the type of food, whether it is high or low acid; the pretreatment; and the method of storage. Water content can be higher in high-acid foods such as fruits, and it should be lower in low-acid foods such as green vegetables, which are more likely to spoil quickly. Meat and fish, which are nonacidic, may contain more water than fruits and vegetables when dried, but the storage life of such foods is also shorter.

"Dried" or "Dehydrated"?

Sources on dried foods frequently confuse the terms "dried" and "dehydrated." The U.S. Department of Agriculture Handbook No. 8, *Composition of Foods: Raw, Processed, and Prepared*, lists dehydrated foods as those containing no more than 2.504 percent water. There is no way that home drying techniques can match this level, so foods prepared at home technically are "dried" rather than "dehydrated."

What Are the Differences among Drying, Smoking, and Curing?

These three terms are also often confused or misunderstood. Foods that are dried have been dehydrated in the sun, shade, an oven, or a homemade or commercial dryer. They often require a pretreatment, one of several processes that varies with the type of food and is used to cut down bacterial growth and spoilage during drying. Fruits, vegetables, herbs, and spices are dried. Meat and fish can also be dried, but this is not recommended for most areas of the United States and Canada.

Smoking is a process of drying food by subjecting it to a high temperature and smoke. Fruits, vegetables, herbs, and spices are rarely smoked, but meat, fish, and poultry frequently are.

Curing is the pretreatment used before smoking.

Who Can Dry Foods at Home?

Anyone can dry foods at home, and almost everyone already has some of the necessary equipment in his or her kitchen. It is easy to start drying foods because small batches can be dried in an oven or even on a tray in a room with a strong breeze and low humidity. When you have discovered the joys of drying and have developed a continuing interest, then you are ready to invest in more elaborate and expensive equipment.

One of the more interesting aspects of home drying is the degree of control you can exert over such things as the foods selected, the equipment, and even the method of drying. Part of the fun of learning to dry foods is experimenting with different methods, and part of the fun of being an experienced drier is the sense of accomplishment that comes from knowing which method can best be put to use on which foods.

Although sun and outdoor drying are not the most highly recommended techniques, by all means experiment with them once or twice. Just remember that the most efficient and effective ways of drying food are usually the indoor methods —using a dryer or a smokehouse.

It is also a good general rule of thumb to start any drying experiments with small amounts of foods. That way, you won't be so worn out from peeling and preparing the food that you can't enjoy the actual drying process—and you can also experiment with several kinds of food, if you like.

Why Bother to Dry Food?

The advantages and pleasures of drying foods are many. Most of the time, it is cheaper to dry food than to purchase equivalent amounts in a store. Dried food is 1/6 to 1/3 the bulk of fresh food and has considerably less water content. It requires less shelf storage space at home, and it is the perfect traveler's food.

Finally, dried food is extremely nutritional. And of course drying is an interesting avocation that can be started with a minimum of mess, equipment, and investment. There is also the somewhat indefinable satisfaction that comes from putting one's own labor into one's pantry—a satisfaction that more and more people have sought in recent years.

Two other interests, in addition to the urge to dry foods, also frequently descend upon the amateur home food drier once he or she becomes a true aficionado. The first is the desire to collect tidbits and folklore related to drying, and the second is the urge to collect dried food recipes. The two go hand in hand since much of the folklore about dried foods includes ancient recipes that, while not always tempting to the palate of today, are nonetheless sometimes interesting to try.

As an example, many food drying fans will enjoy knowing that dried snake was a favorite travel food of the Chinese more than 2,000 years ago. Few will actually want to test the recipe. Another interesting piece of dried food miscellany: the famous Bombay duck was actually a dried fish carried by traveling Asian Indians almost 1,000 years ago. Recipes for charqui and pemmican, two other dried food originals contributed by American Indians, are discussed in detail later in this book.

It is even possible to duplicate a form of dried vinegar that travelers can conveniently carry in their pockets. The technique, which comes from a seventeenth-century cookery book, is fairly simple: "Take the blades . . . of either Wheat or Rye, and beat it in a morter with the strongest Vinegar you can get till it comes to a paste; then roll it into little balls, and dry it in the sun until it be very hard. Then when you have occasion to use it, cut a little piece thereof and dissolve it in wine, and it will make a strong Vinegar." It does indeed make the perfect portable condiment.

Other more useful recipes are scattered throughout this book, and you will surely want to begin your own collection to go along with your growing interest in drying foods.

2

THE EQUIPMENT
ESSENTIALS

THE EQUIPMENT
ESSENTIALS

The cost of home drying equipment ranges from a few dollars for wood to build a solar dryer to several hundred dollars for an elaborate commercial dryer. Fortunately, the amount of money spent on drying equipment has little to do with the results. Even the most elaborate dryer is only a box to hold the food, since good drying results from high-quality food, proper drying techniques, and maintenance of a temperature that is neither too high nor too low.

In addition, it is silly to acquire expensive and elaborate equipment to dry small batches of food. If you plan to dry 5 or 6 pounds of some relatively inexpensive vegetables once a year, for example, you can do the job most efficiently and inexpensively in a regular oven.

Although building your own equipment—and somehow it is more challenging to build your own—does not require a large monetary investment, it does mean an investment in time and energy. For this reason, it is a good idea to be sure you enjoy drying food before making any major commitments in money or time. Start by drying small batches of food in an oven or outdoors. If you find that the task proves to be enjoyable, it is time to start collecting and building your own equipment.

Basic Drying Equipment

What exactly is necessary equipment for drying food? Very little, as the list that follows shows. In addition to trays and a dryer for indoor use, several other pieces of equipment are especially helpful and necessary. There is no reason to run to the nearest store and buy everything listed below. Borrow, use things you already own—in short, improvise, an act that is at the very heart of the food-drying process anyway.

Wood blocks or cement blocks. These are used to lift the bottom tray away from the ground and dampness, to separate the tray from the burning sulfur during sulfuring, and to separate stacked trays in a dryer. Any scrap material that works can be used for this.

Sulfuring materials. Some fruits dry better if they are sulfured first, and a few simple materials are required for this process. These are described in detail in Chapter 4, but the basics are a cardboard box, sulfur, a container for the sulfur, and trays.

Thermometer. An oven or meat thermometer works for dryers. Just be sure it is accurate, since the range of drying temperatures does not vary much, and a thermometer that is off by only 10 degrees could cause food to cook or scorch rather than to dry.

Dryer. A dryer, also called a dehydrator, is what most people develop an urge to build after they have discovered the pleasures of drying food. A dryer can be as simple as your own oven, turned to an appropriately low temperature and loaded with food-laden trays. It can be the sun, if you live in the right part of the country. But here the term refers to a homemade dryer. Detailed instructions on how to build several types of homemade dryers appear later in this chapter.

Knives, shredders, and a grater. Food is frequently peeled, cut, shredded, or grated prior to drying. Any equipment that helps to speed this preparatory process is desirable.

Several large pots and bowls. One very large pot is required for steaming and blanching fruits and vegetables prior to drying. The cheapest and most convenient solution is to purchase one of the black-speckled canning kettles that are sold in home centers and hardware stores. These kettles generally retail for under $10, a good price compared to the cost of most lines of cookware. In addition, a steaming rack, basket, large strainer, or colander that fits comfortably inside a smaller pot is useful for steaming. Other pots or bowls for holding food during an ascorbic acid bath or other preparations are also helpful.

Scale. This is not a must if you are just starting out, but it is useful in weighing food to help judge how much moisture has been lost and thus to test for dryness. It also helps to determine the amount of sulfur required for sulfuring some fruits. A scale is not necessary if you know the weight of food when you buy it.

Electric fan. This is useful for speeding up shade drying and room drying. Be sure to buy a fan with a protective covering over the blades. Fans are not recommended for use during oven drying—it is too risky to operate a piece of electrical equipment near a gas or electric stove.

Cheesecloth or muslin. Many yards of either of these materials are needed to line and cover trays that are used outdoors.

Tarpaulin or canvas. You might or might not need this item, depending upon whether or not you are drying food outdoors and how low the temperature goes at night.

Cleaning materials. A strong detergent and a stiff brush are necessities for cleaning your equipment. Trays especially should be thoroughly cleaned after use.

Building Drying Trays

Trays are the most frequently used type of drying equipment. They are also generally the first equipment acquired. The first and most important factor in building trays is to select a size. A tray should not be too large to be comfortably lifted by one person when it is full of food. Since 1 square foot of tray can hold 1½ to 2½ pounds of food, an average-sized tray weighs about 5 to 8 pounds when fully loaded. Generally, trays become awkward to handle when they are more than 2 feet in length and 12 to 18 inches wide.

Even if you are planning to use trays in your own oven or outdoors, give some thought to planning tray size with an eye toward eventual use in a homemade dryer. Trays should be 3 inches shorter all the way around than the dryer's internal measurements, to allow for maximum air circulation.

Two inches is the ideal tray depth, for several reasons. For one, it is the best depth for sun-drying. Also, this depth is perfect for use in an indoor dryer because it creates the space needed between trays for air circulation.

Choosing the Best Tray Materials

Wood is the single best material for a drying tray. Metal is corrosive and may react chemically with food, so it is generally ruled out as a tray material. Any lightweight, durable wood will suffice, with a few exceptions. Green wood should not be used because of a tendency to warp. Pine is resinous and should not be used with food because of the fumes it can impart to it. Oak and redwood have a tendency to stain. The wood for trays can be salvaged from almost any source—an old packing crate, for example—or it can be purchased new.

Years ago, galvanized screen was considered a good material to use on the bottom of trays. Some sources still recommend its use, but galvanized screen is coated with zinc and sometimes contains cadmium, two materials that are best avoided around food. Copper screening is also unsatisfactory, because it destroys vitamin C in food. Vinyl-coated screening, the newest addition to the marketplace, does not work well at high temperatures because it has a tendency to melt. On the other hand, it is satisfactory for outdoor drying.

With the exception of these screening materials, you can use anything on the bottom of trays that can be adequately strung across them. Half-inch wood strips or dowels are probably the best material to use; twine or lightweight wire has been known to do the trick. The slats from wooden packing crates are ideal. Whatever is used, it is a good idea to cover the bottom of the tray with a single layer of cheesecloth or muslin to prevent staining or sticking. A clean cloth should be used each time; the cloth should be laundered between uses. A light coating of mineral oil also works well and makes the trays easier to clean. The tops of the trays are left open.

How to Build the Trays

A good standardized size for trays is 12 by 22 by 2 inches; it will fit home-built dryers and your own oven. To build one tray, you will need a piece of wood ⅛ to ¼ inch thick and measuring 8 by 24 inches. Cut two short, end pieces 12 inches wide and 2 inches deep. If you have bought wood 24 inches long, you can cut one long piece and divide it in half to get the two short pieces. The long tray sides should be 22 by 2 inches. Then cut three 12-by-1-inch pieces; again, these can be cut from one piece and split in half.

If you are using dowels or slats made of wood, cut these 12 inches long; the number required will depend upon the material you are using and how far apart you plan to place them. Screening or other similar material used for the tray bottom should be a rectangular sheet 12 by 24 inches.

Begin by cutting the wood to the size you have chosen. Nail together the sides. You may want to use cleats at the corners to brace the wood and make the trays sturdier. Attach the dowels or wood strips to the sides, making them no more than ¼ to ½ inch apart. (It is a good idea to make one or two trays with ¼-inch spaces on the bottom and several others with ½-inch spaces; this way you will have trays to accommodate any size food.)

Nail three 1-by-1-inch wood strips across the short ends and in the middle of the tray (see Figure 1A). They should be extensions of the ends. This prevents the wood from warping as it is subjected to heat, and makes the trays sturdier. And, if you have built trays 2 inches deep, it will also create the space necessary to separate the stacked trays three inches for perfect drying.

As a finishing touch, paint numbers consecutively (1 through 6, or whatever) on one end of each tray. Trays are rotated frequently to insure even drying, and this will help you keep track.

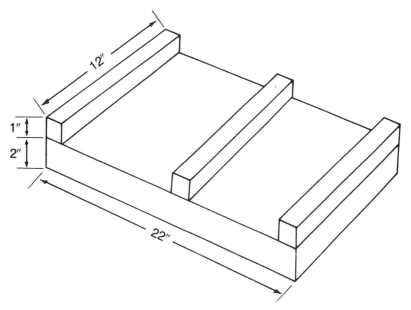

Figure 1A. Bottom of tray

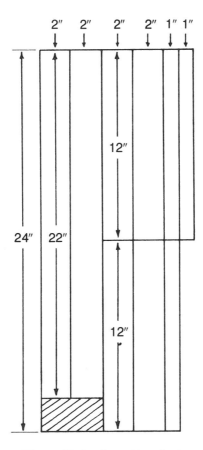

Figure 1B. Cutting pattern for tray

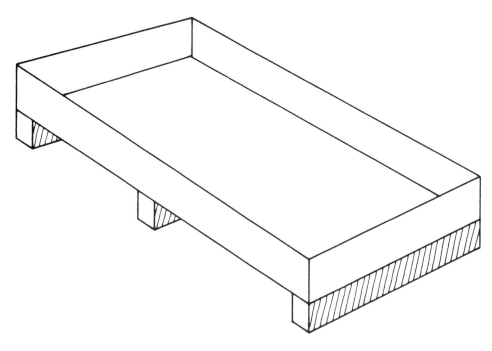

Figure 2. Top of tray

Cleaning and Storing Trays

Trays should be thoroughly scrubbed with detergent and a stiff brush after each use. They can be stored upright or flat in any convenient place that does not have extremes of temperature.

Solar Dryer

A solar dryer is excellent for use in drying small quantities of fruits and vegetables outdoors. A small home unit can easily be built utilizing a window frame.

The principle of a solar dryer is similar to that of a forcing box for starting plants. The rays of the sun heat the food under the glass with a far greater intensity than they would if there were no glass, since less heat is lost to the surrounding air in such a container. This results in a higher temperature within the dryer than in the surrounding outside air. Because of this, a solar dryer works especially well in climates where the air is not particularly hot or dry.

Basically, a solar dryer is a container of wood (or any other material suitable for outdoor wear and tear) that is flat on the bottom, back, and sides and slanted on the top, where the window frame or other glass panel is used. The sides, front, and top back should have ventilator spaces to allow a free flow of air, as shown in Figure 3.

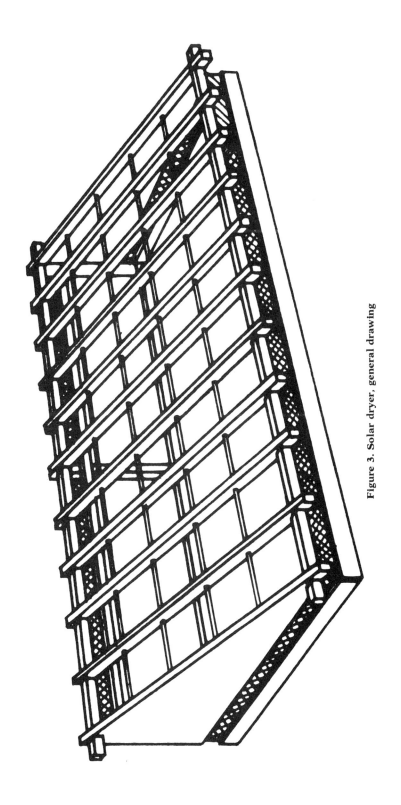

Figure 3. Solar dryer, general drawing

These ventilators should be covered with galvanized wire mesh or mosquito netting—something thin-meshed enough to keep out insects.

Sample measurements are given in Figure 4, but the size can easily be varied to suit individual requirements.

In addition to the wood and netting, you will need nails, 4 hinges, and 4 fasteners.

Here is the basic procedure for building a solar dryer:

Lay the bottom or base (floor) board on a level surface. Select one 4-inch-wide strip of wood that has been cut exactly to the length of the long (A) sides of the base. Nail or glue this in place at what will be the rear (high) end of the drying box. Select two more 4-inch-wide pieces of wood and nail or glue these along the short sides of the bottom board, level with the wood strip attached to the back.

Attach the two doors with hinges to the bottom strip of wood at the back. These are the openings through which the food is passed. At the top of the doors, attach the top fasteners to a solid strip of wood (B), as shown in Figure 4. Attach the sides (C), positioning them 6 inches above the side wood strips (D).

Put the window frame or whatever you are using as the glass side in place, but do not attach it yet. Carefully attach the second long piece of wood to the front of the box. Rest the window frame on top and test to make sure that it and the strip of wood fit together perfectly. If they do not fit smoothly, saw or sand part of the window frame or the wood strip to make them fit. Remove the window frame. Tack or nail the cloth or netting over the vents on the front or sides. Attach the window frame permanently to the top of the box, using a strip of wood to join the frame to the body of the box. Flip the box over; open the hinged doors and nail or tack in the last piece of netting, which goes between the window frame and doors.

Sand all the surfaces with which the window frame has contact so that it fits tightly against all surfaces. Make the dryer watertight by sealing with putty those edges that are not meant to open. This will also help prevent insects from getting in the dryer. Food can be placed on trays, paper towels, or cheesecloth in the bottom of the dryer.

A variation in the solar dryer consists of leaving the glass panel unattached so it can be used as the door for inserting and removing food. This is practical only for very small dryers.

If the dryer is carefully checked for watertightness before each use, it can be stored outdoors. Simply drape a canvas or tarpaulin over it for protection when not in use. Indoors, store it in a garage, basement, closet, or other out-of-the-way place. When the dryer has not been used for a long time, check it carefully to be sure it is watertight before using it again.

This type of solar dryer can be built larger to accommodate greater quantities of fruits and vegetables. It is best, however, to extend the length and not build a solar dryer any wider than 4 feet. Any larger size makes it awkward to take food in and out.

Small dryers should be shifted several times a day to accommodate the sun's rays to the fullest extent. This is impossible to do with larger solar dryers, so these must be built to accommodate the greatest amount of sunshine available in any climate.

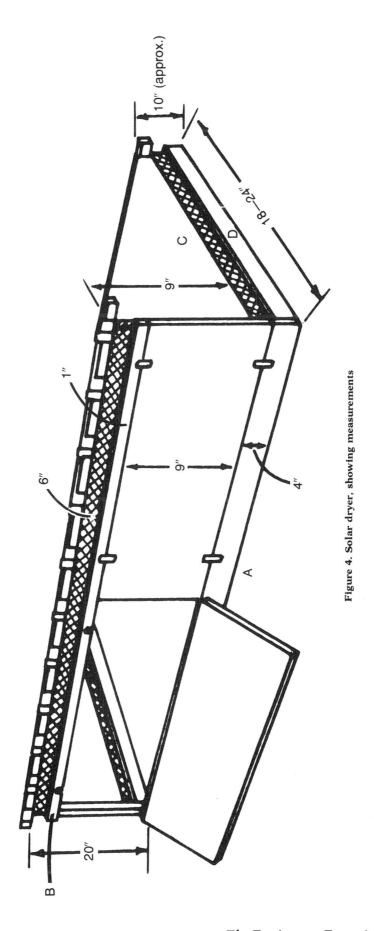

Figure 4. Solar dryer, showing measurements

Stove-Top Dryers

There are two kinds of stove-top dryers that are easy to build and convenient to use. When the wood-burning stove was replaced by electric and gas stoves, these dryers, which worked especially well over the old-fashioned stoves, fell into disuse. Stove-top dryers adapted to today's stoves, however, have reappeared as the interest in putting up food at home has grown. These dryers can be used on gas, electric, or wood-burning stoves.

In addition to being easy to build, stove-top dryers are convenient for drying small amounts of food. One variety, the birdcage dryer, is somewhat awkward on occasion as it is meant to be mounted permanently on a kitchen wall. If this presents a problem, the cabinet dryer is easier to store away and is less awkward to shift around during drying. But both models dry food equally well and can be built to any size.

The birdcage dryer, shown in Figure 5, is designed to handle three stationary trays. Dryers with more than this number of trays become awkward to handle because of their weight.

Before building a birdcage dryer, investigate your kitchen to be sure that the wall behind the stove is solid enough to support the dryer's weight. If not, the dryer can also be suspended from the ceiling. Basic equipment needed for the birdcage dryer includes:

- 21 square feet of aluminum screen or baling wire
- 7 square feet of muslin
- 75 feet of 1½-inch-wide wood strips
- steel brace strong enough to hold the dryer
- wire heavy enough to support the weight of the dryer
- screws, nails, and tacks

Cut the wood strips into 8 pieces, each 2 feet long. These are the vertical braces.

Cut the remaining wood strips into 16 pieces, each measuring 2 feet, 6 inches long. These pieces form the square frame, the outline of the "trays." Using nails, join 4 of the 2 feet, 6 inch pieces to make a square. Repeat the process 3 more times. One by one, align the vertical wood braces inside the frames as shown in the drawing. Leave approximately 6 inches between each drying area.

Nail the vertical wood braces in place. On one end of the frame, nail the muslin in place carefully. This is the top of the dryer. On the 3 remaining "tray" frames, tack down the screening or string the baling wire in a crisscross pattern. Sand any rough edges.

Mount the wood strip on the wall behind the stove, using the steel brace for support. Tack the wire to the 4 corners of the dryer; it will be used to hang the dryer from the support. Adjust the length of the wire so the birdcage dryer will be approximately 10 inches from the source of heat when in use.

Don't be tempted to use muslin or cheesecloth on the tray areas; these materials catch fire too easily.

To use the birdcage dryer, arrange the food on the trays. Lower the dryer to a point about 10 inches from the heated burners. Clip a thermometer somewhere near the top tray. Dry food according to directions for each variety.

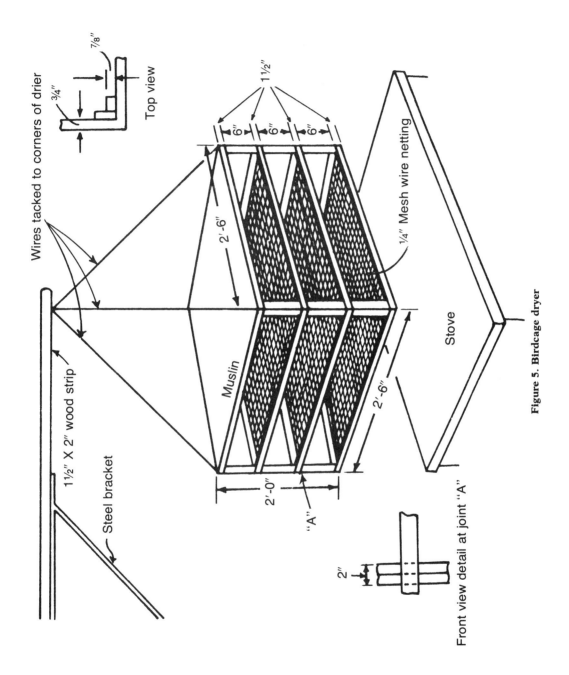

Figure 5. Birdcage dryer

Top view

⁷⁄₈"

¾"

Wires tacked to corners of drier

1½" X 2" wood strip

Steel bracket

Muslin

2'-6"

1½"

6" 6" 6"

¼" Mesh wire netting

Stove

2'-6"

2'-0"

"A"

Front view detail at joint "A"

2"

Building a Cabinet Dryer

One of the more modest dryers that is nonetheless excellent for people who are not yet ready to plunge into drying with more elaborate equipment is the simple cabinet dryer.

It should be built more or less to the proportions of your stove. The trays suitable for use in this kind of dryer were discussed earlier in the chapter. Their bottoms must be wood or nongalvanized screen; cloth is too likely to catch fire. The trays should be about 3 inches shorter than the depth of the dryer so they can be staggered for quicker drying. Such a dryer can be used over a hotplate as well as a stove. This dryer, which has a capacity for 6 shelves, can dry 20 to 30 pounds of food at a time.

Build the dryer of a nonresinous hardwood. Wood is needed for the sides and the door. It is optional on the top, which you may prefer to cover with muslin or screen for better air circulation. There is no problem in using nongalvanized or galvanized screen where it does not come into contact with food. You will need two hinges for the door, and twelve 2-by-1 strips of wood for supports for the trays. If the top is not left open, side vents must be built in. A metal heat deflector is used in the bottom; it should be about 2 inches shorter all the way around than the outer dimensions of the box. Aluminum, or even an old cookie sheet or other piece of scrap metal, works well here.

Begin construction by cutting all the wood to the sizes desired. Nail the tray supports, 4 inches apart, to the sides of the box. Put the frame together. Install the heat deflector.

Use long nails or screws, attached at right angles to the sides of the dryer, to support the heat deflector. Open space around the heat deflector permits better air circulation.

The legs of the dryer must be nonflammable. They can be made of the same wood used on the base, covered with metal, or long screws or nails can serve as legs.

To use this dryer, simply place it on top of the heated burners of a stove. The heat deflector, and therefore the bottom of the dryer, should be 6 inches from the source of heat. Clip a thermometer near the top tray. Watch carefully to regulate heat. Place the trays inside, taking care to push every other tray as far to the back as it will go and to bring alternate trays as far to the front as they will come.

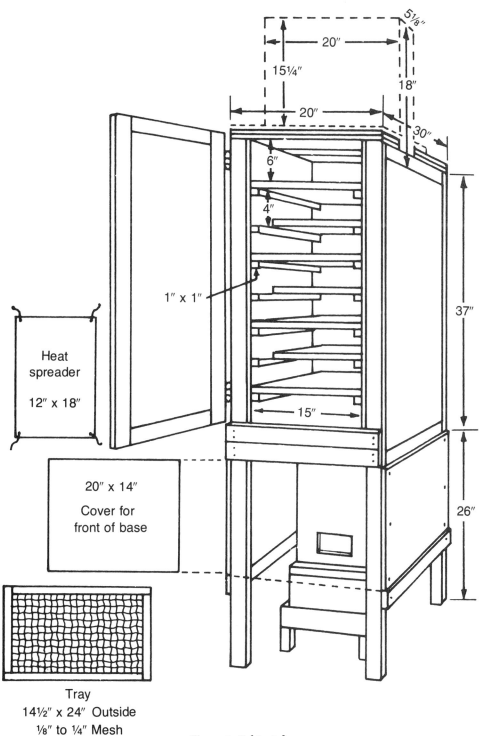

5⅛"

20"

15¼"

18"

20"

30"

6"

4"

1" x 1"

37"

Heat
spreader

12" x 18"

15"

20" x 14"

Cover for
front of base

26"

Tray
14½" x 24" Outside
⅛" to ¼" Mesh

Figure 6. Cabinet dryer

Portable Home Dryer

The most efficient home dryer, easily built for home use, is the one with its own independent heat source. It is portable and can be used anywhere there is an electrical outlet. The ideal location is an out-of-the-way place, such as a porch or hallway, where it is not an obstruction and can be checked frequently.

A portable home dryer represents an investment of $80 to $120, depending upon the materials selected and their cost locally. It will dry approximately 30 to 50 pounds of food in each batch, and can easily be built by someone with little experience in carpentry. Even the heating unit is fairly simple; it can consist of a hotplate, light bulbs, or any other heating methods you may devise.

This dryer can be made of any nonresinous wood.

Very little special equipment is needed for this version.

A thermometer, permanently mounted, that covers a temperature range from 100° F. to 180° F., is helpful.

Trays are needed, of course, to hold the food. This dryer is designed to hold no more than 7 trays at a time.

The heating apparatus can be a hotplate, light bulbs, a small floor heater —almost anything with 600 to 1,000 wattage. (The size of the heating unit varies according to the size of the dryer; if you use smaller dimensions than are listed here, the wattage of the heater should be appropriately reduced.) The safest, most easily purchased, and most convenient heating unit to install consists of **light bulbs** (75 watts each), which are placed in porcelain sockets to further protect them from the heat generated during drying.

An electric fan is required to circulate the air throughout the dryer. Make sure that it reaches all trays more or less equally. A good size is a 1/150 horsepower fan with 3-inch-diameter blades.

The body of the dryer is similar to that of the stove-top cabinet dryer. It is basically a box container, with the heater and the blower below.

Figure 7 shows how the entire dryer can be cut from one piece of ½-inch thick plywood.

In addition, other required materials are:
•24 feet of 1-by-1-inch wood strips (tray supports)
•2 hinges for door
•1 latch for door
•9 porcelain surface-mount light sockets
•9 75-watt light bulbs
• thermostat, waterproof or immersion type, with a temperature range of 100° to 160° F.
• wood screws and nails
•aluminum or other metal piece 16 inches square

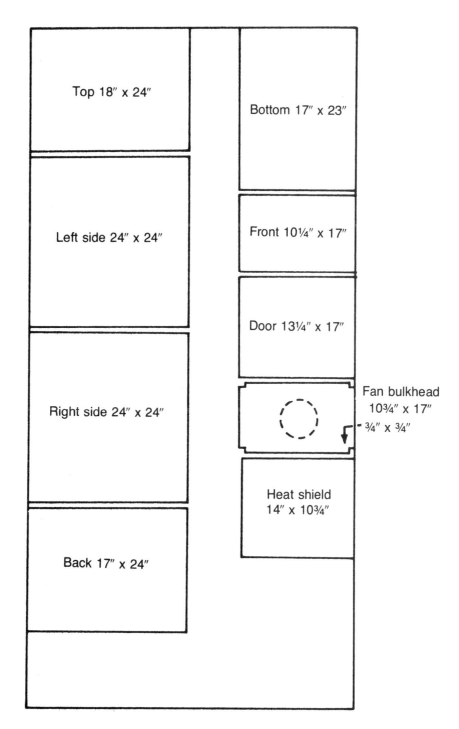

Figure 7. Cutting pattern for homemade dryer

Equipment needed is equally simple: a saw, drill, screwdriver, measuring device, and a countersink. The most complicated part of the construction—attaching the fan and heater to the thermostat, and installing an on-off switch—really should be done by an electrician. The job is fairly small, but requires professional expertise unless you are an experienced amateur.

To construct this dryer, cut out the pieces as shown in Figure 7. Cut the wood strips into pieces 18 inches long. Attach the tray supports (wood strips) to the side panels. Start at the top and place the first tray supports 3 inches from the top of the wood panels. Measure off 4 inches and attach the second set of tray supports. Continue in the same manner until all 7 sets of tray supports are in place, making sure to leave 4 inches between each set. Attach the metal heat shield 3 inches below the last (bottom) set of tray supports. Do this by resting the shield on nails which have been driven through the sides of the dryer at right angles.

Attach the left side to the base, supporting it with a 1-by-1-inch strip, as shown in Figure 8. Drill an air hole 1½ inches wide in the lower front section. While you are working with the drill, make two or three additional air holes toward the top of the door; then put the door aside. Attach the lower front section to the base and the left side, then arrange the porcelain sockets as shown in Figure 8. Finally, fasten the sockets in place with screws.

The electrical work is the next stage: Figure 9 shows how it should be set up. Basically, asbestos-covered copper wire is attached to yellow screws on the light bulb sockets and used as conductor between the light bulbs and the thermostat. The fan, which should be secured to the baseboard along with the porcelain sockets (see Figure 9 for exact position), is also connected to the thermostat via an extension cord wire. A utility box can be used to join all the connections, and the wires should be fastened securely to it. It may also be necessary to build a partition between the fan and the heater, depending upon the type of fan used.

Before attaching the front door, finish the inside of the dryer. It may be painted with aluminum paint or covered with aluminum foil. Both will help to deflect the heat from the wood onto the food.

Install the right side, reinforcing it at the base joint with a wood strip. Attach the front door, using the two hinges on whichever side is most convenient for you.

Finally, sand any rough outside edges and, if you wish, paint the outside of the dryer. Before drying any food in the dryer, run it for half a day, both to check the thermostat for accuracy (use a meat thermometer inside to read the temperatures) and to rid the container of any wood or other miscellaneous odors.

To dry food, start the engine (fan and heater) and bring the dryer to the desired temperature recommended for whatever food is about to be dried. Place the loaded trays in the dryer one by one. Push the first tray against the back wall of the dryer; pull the second tray to the front door of the dryer. Continue alternating trays in this manner until the dryer is loaded. As a general rule, start drying with the door latched. As drying progresses, and depending upon the type of food, it may be necessary to open the door slightly to adjust the temperature and increase the flow of air. If the door tends to swing farther open than desired, use a piece of wire to hold it in place.

Detailed instructions for drying individual fruits and vegetables are discussed in Chapter 4.

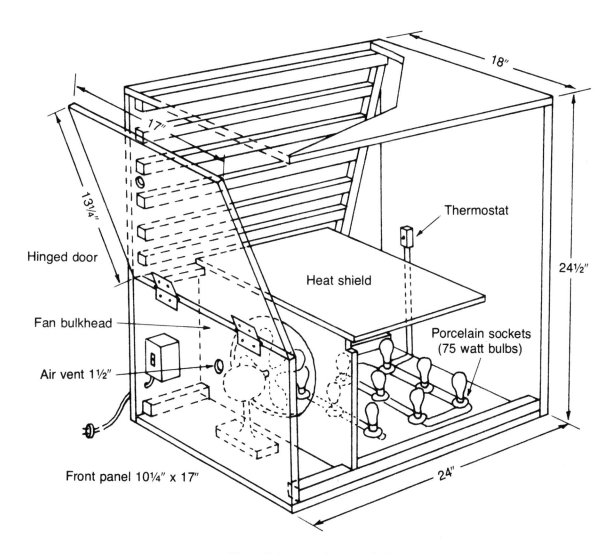

Figure 8. Cutaway, homemade dryer

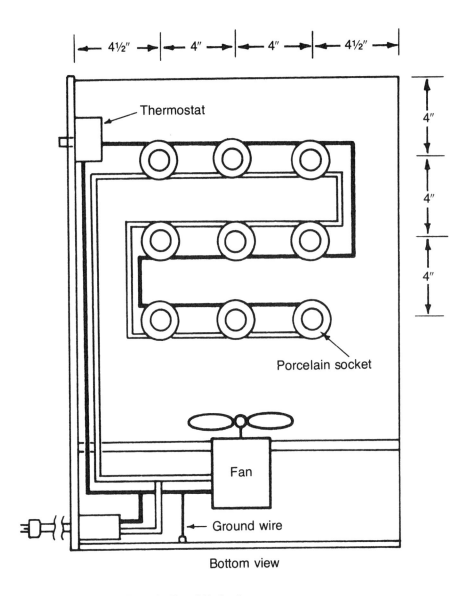

Bottom view

Portable Electric Food Dehydrator

Layout of socket locations and wiring plan
for dehydrator with thermostat.

Figure 9. Porcelain socket arrangement

3

DRYING HERBS

DRYING HERBS

Herbs and spices add a final, yet subtle touch to plain and fancy cooking and are easy to dry. The beginner seeking encouragement will find it here; the methods are simple and the results are sure.

First, it is useful to make the distinction between herbs and spices. Herbs are the leaves, and occasionally the seeds and flowers, of aromatic plants. Spices are the parts of these plants that may be harvested for food—such things as the buds, bark, roots, fruits, and seeds.

Finding Herbs to Dry

If you have access to a wooded area, you will find gathering wild herbs a tremendously enjoyable experience. Most libraries have books on local plants that will help you get started. Another ideal source of herbs and spices is to grow your own herb garden. Aromatic plants are varied and hardy for the most part, and many people have begun to discover the joys of cultivating their own herbs and spices. Some herbs take quite well to sunny city windowsills, and many can be grown under special plant lights.

If you can't gather wild plants or supply yourself from your own garden, try a greengrocer or the produce department of a large grocery store. The only drawback to buying plants here is that they often have been cut several days earlier; thus you are dependent upon what happens to come into the store. Finally, arrangements can probably be made to order aromatic plants from a reliable plant store. Whether buying from a greengrocer, store, a plant dealer, or gathering herbs from the wilds, take care to select the most perfect specimens you can find.

Choosing the Herbs and Spices to Dry

Almost all herbs and spices can be successfully dried. The exceptions are chervil, chives, parsley, and basil, which in drying do lose some of the flavor for which they are cherished when fresh. This is no reason, however, to rule them out entirely as dried foods; it simply means a little more thought must be put into their use. Dried chives, for example, work fairly well in moist foods, such as cottage cheese and dips. Dried basil is acceptable in tomato sauces and soups. It is better, however, to avoid using it in salads, egg dishes, or in pesto, that marvelous basil-based sauce, if it cannot be obtained fresh. While unable to stand the test of use alone, dried parsley, basil, chives, and chervil can be successfully used in combination with other herbs. Their flavor will never be as complete or as subtle as when fresh, but it will suffice during those months when fresh herbs simply are not available.

Drying herbs and spices should also increase your interest in using them. Be on the lookout for unusual herbs and spices and for out-of-the-ordinary ways to utilize them. Finally, don't overlook the possibilities of drying such plants as angelica, burdock, comfrey, rose petals, ginseng, and sassafras—plants that are not frequently used in cooking, but which make excellent teas and interesting sachets and potpourris in dried form.

Harvesting Herbs and Spices

The best time to harvest herbs and spices is in the early morning, just after the dew has burned off them. Simply snip off 5 or 6 inches of top growth with gardening shears or scissors. Leafy herbs—such as mint, rosemary, tarragon, or lemon balm—should be cut just before they bloom and when their leaves have reached mature size. This is when the oils that provide the essence of the herbs are at their peak. Most leafy herbs can be harvested as many as three or four times during the summer. The same is true for flowering herbs. Seeds, on the other hand, should be harvested in the fall, just after their color has changed from green to brown or gray. Flowers should be harvested the day they bloom, while roots are best dug up in the fall or early winter, when they are going into their dormant stage and their flavor is at its peak.

Preparations for Drying

Freshly harvested aromatic plants should be washed in cold water, shaken free of excess water, and tied in small bunches. They can then be hung in a warm, sunny place until the water has evaporated. With careful supervision, herbs and spices can be completely dried in direct sunlight, but they are generally best when dried indoors or in a shady place.

A hundred years ago, herbs and spices were tied in small bunches or strung over an open hearth. Later, they were dried over a wood-burning stove or hung in a warm attic. Today, they may be dried just as easily by hanging in a warm, dry place in the kitchen.

Methods of Drying

There are several simple methods of drying herbs and spices at home. Each method requires that they be dried quickly (generally 3 to 5 days) in order to preserve the fullest possible aroma and essence. Room-drying is the simplest way. To do this, place the herbs on trays lined with cheesecloth or fine mesh and leave them to dry. If they do not fully dry within 10 days, they should be helped along with oven-drying. This consists of placing the trays in a 100°F. oven and letting the contents dry for several hours. Keep the oven door ajar as a means of regulating the heat. Check the trays every ½ hour to prevent scorching. Herbs are fully dried when they crumble when rubbed between the fingers.

Herbs can also be tied in small bundles and hung upside down in a warm room. (This is done to permit the oils to concentrate in the leaves.) For added protection, especially from light, which tends to discolor herbs and spices, they can be enclosed in a small brown paper bag. This method also protects the herbs from absorbing cooking grease and dirt in the air. Holes of about ¼ to ½ inch in diameter should be punched in the bags to permit air to circulate throughout. This protection is especially helpful in drying sage, savory, oregano, lemon balm, and mint.

Herb leaves can also be dried on trays in the oven or in a dryer. Maintain the temperature at about 100°F. and circulate the trays every half hour so that the herbs in the top and bottom trays do not scorch or dry more rapidly than the others.

Figure 10. Brown paper bag container for drying herbs

Seed pods can simply be placed on a tray or cookie pan and left in a sunny window to dry. To release the seeds, just rub the pods between the palms of the hands after they are dried. Alternately, seed pods can be dried by hanging them upside down in small brown paper bags. As the pods dry, the seeds will fall to the bottom of the bags.

Flower petals can be dried on a tray in a warm room. Just separate the petals from any protective covering (the calyx) and trim away any tough portions on the inner tips. Rose petals particularly need this treatment. Flower heads that are to be used for tea, such as camomile, should be dried whole.

Herb roots should be carefully washed in cold or lukewarm water, then shaken or towel-dried to remove excess water. They will dry in a few days on a tray in a warm room.

After leafy herbs have dried, you may want to separate the leaves from the stems. If so, do not break or crumble the leaves until you are ready to cook with the herb, as this releases their essence.

A simple and attractive drying device for herbs and spices can be made using a ¾-inch dowel rod and a few screw hooks of the type frequently used to hang plants (see Figure 11). Screw the hooks into a ceiling, equidistant from each other and less than the length of the rod. Use string to tie fresh herbs and spices (in brown paper bags or not) to the rod. The rod could be painted, or you might find it more attractive unpainted. When it is not being used to dry plants, the rod can double as a pot holder or hanging plant rack.

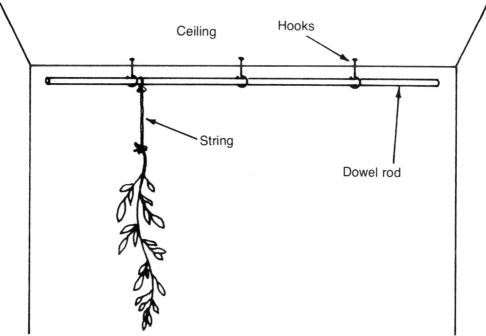

Figure 11. Dowel hanger for drying herbs

Conditioning

Like fruits and vegetables, herbs and spices need about a week of conditioning before they can be permanently stored away. Place each batch of dried herbs or spices in its own tightly sealed jar. Unseal and gently shake or rearrange the plants several times during the week. If moisture collects inside the containers, spread the plants on a tray and oven-dry until crumbly. Mold is likely to develop if this step is skipped.

Storing Herbs and Spices

Properly stored, herbs and spices will keep about one year, sometimes longer. The test is smell—if they still smell strong, they are usable. Herbs and spices should be stored in tightly closed, opaque containers and kept in a dark place. They should not be stored in cardboard containers, which tend to absorb their oils.

Using Dried Herbs and Spices in Cooking

There is only one rule of thumb in using dried herbs and spices: they are stronger than fresh ones and a little goes a long way. Use about ⅔ less of dried herbs and about ¾ less of powdered spices or herbs than you would of the fresh varieties.

Some people feel that dried herbs used in cold foods should be rejuvenated with a presoaking for several hours prior to use. This is really a matter of taste, so experiment to see what works best for you. After you have dried a selection of herbs and spices, consider mixing your own *fines herbes, bouquets garnis,* and *épices fines* for later use. Despite their French names and origins, these are simple combinations of spices and herbs. No recipe is needed for a *fines herbes* mixture; simply combine what smells and tastes good to you. Often *fines herbes* are a mixture of chives, chervil, tarragon or rosemary, and parsley. This mixture is kept near the stove in most French homes and is used to season a great variety of dishes. Although *fines herbes* are at their best when fresh herbs are used, premixed combinations of dried herbs are quite good too. Even more adaptable to dried herbs and spices is a Mediterranean *fines herbes* mixture, which generally includes oregano, basil, sage, saffron, and sometimes fennel. *Fines herbes* can be used in many imaginative ways: in casseroles, hamburger dishes, meat loaf, on roasts, in salads, and with egg dishes.

A *bouquet garni* is usually a combination of thyme, bay leaf, and parsley tied together in a cheesecloth. It is used to season soups and stews, as well as stocks. Dried celery is also a frequent addition to *bouquet garni.*

Epices fines refers to a cook's own combination of spices. It is used in pâté, meat loaf, marinades for meat, and homemade sausages. A simple way to make a delicious breakfast sausage is to combine 2/3 of a tablespoon of your own *épices fines* with about 2 pounds of high-fat ground pork. In making sausages and meat loaf, some cooks use a combination of *épices fines* and *fines herbes. Epices fines* can be purchased prebottled (it is often called *quatre épices*), but it is much more interesting to make your own, particularly if you are using spices and herbs you have dried yourself. The quantities can be varied to suit personal tastes, but this is a basic recipe for a highly spicy *épices fines:*

½ cup black peppercorns	*1½ Tbs. thyme*
4 bay leaves	*2 tsp. basil*
1 Tbs. powdered or whole clove	*1½ tsp. cinnamon*
1 Tbs. nutmeg	*1½ tsp. marjoram*
1 Tbs. Hungarian or Spanish paprika	*1½ tsp. sage*

Mix all ingredients together in a blender container and blend until pulverized. Store in a tightly sealed jar or metal container away from light.
Use 1 to 1½ teaspoons for every 4 cups of meat, or experiment to taste.

By keeping mixtures of *fines herbs, bouquets garnis,* and *épices fines* at the ready, you will soon find yourself experimenting, with the result that you will uncover almost endless uses and subtle variations for dried herbs and spices.

Herb butters, which may be used as toppings for broiled meat and fish, as poultry basters, and in sauces, are another interesting way to use dried herbs and spices. They are also delicious on warm breads and in egg dishes. The technique for making all herb butters is basically the same:

Let ½ cup butter soften. Blend in one tablespoon of an herb, such as rosemary, tarragon, parsley, chives, or *fines herbes*. One tablespoon of lemon juice or meat glaze may be mixed in to add additional flavor.

Finally, for anyone who has ever developed a taste for the scent and flavor of fine Creole cooking, there is one vital ingredient that is not always readily available in stores, but which can easily be made at home: filé. Filé is the seasoning and thickening agent that imparts such a unique flavor to gumbo, and it consists of nothing more than dried and powdered sassafras leaves. Here is an authentic gumbo recipe to go along with the filé you can make:

DELICIOUS OLD-FASHIONED GUMBO

4 Tbs. vegetable oil	6 cups boiling water
1 onion, chopped	1 bay leaf
1 cup ham, diced	¼ cup fresh chopped parsley
4 cups cooked chicken,	½ tsp. dried thyme
chopped into bite-sized pieces	¼ tsp. cayenne powder
2 1-lb. cans tomatoes	1 tsp. filé
2 10-ounce pkgs. frozen okra	4 cups hot cooked rice

Heat the vegetable oil in a large pot, and sauté the onion and ham in it until the onion is limp. Add the chicken and stir-fry for 1 minute. Add the tomatoes, okra, boiling water, bay leaf, parsley, thyme, and cayenne. Cook uncovered for 10 minutes, stirring occasionally and using a wooden spoon to break up the tomatoes. Cover and cook at a simmer for 45 minutes. Remove from heat and stir in the filé just before serving. Do not let gumbo boil after filé has been added or it will become stringy. Put about ½ cup of rice in each soup bowl and ladle the gumbo over it. Serve at once. Serves 8.

Usual and Unusual Herbs and Spices

Here is a list of herbs and spices that can be grown or gathered and dried, along with special drying hints and general uses for each variety. *Note:* These herbs and spices have many other uses, of course, when they are fresh. These descriptions are limited to their uses when dried.

angelica Seed gathered in fall. Used in cake or cookie dough.

basil Leaves are gathered throughout summer when plants are about to bloom. Basil and tomato have an affinity for each other; basil is also good in egg dishes, soups, and with fish and shellfish.

bay This herb is grown only in warm climates, where the mature leaves may be harvested any time. It is used in stews, soups, roasts, and in tomato and meat sauces.

borage Leaves are harvested when plants are mature. Flowers are gathered on the day they bloom. Leaves and flowers are used in salads and on cooked vegetables.

burnet Dried leaves are gathered in the fall. The faint cucumber flavor of this herb enhances cooked vegetable dishes and makes a soothing herbal tea.

camomile The flower is gathered the day it blooms. Camomile petals make an excellent herbal tea.

caraway Seeds are dried, but the entire plant must be cut as soon as the seed pods have ripened because the seeds disperse very quickly. Dry the entire plant on paper or tray. Used in Hungarian cookery, pork dishes, breads, baked desserts, and cheese dips.

celery Leaves are dried when mature. They are used to season soups and are often a main ingredient in *bouquets garnis*. Seeds, which are harvested in fall, are used in coleslaw, pickles, and relishes.

chervil Harvest leaves when they are still young. Dried leaves may be used in soup, egg dishes, celery and cucumber dishes.

chives Can be harvested throughout their growing season. Trim about 1½ to 2 inches from the roots. Chives add gentle onion flavor to cottage cheese, egg dishes, and cream sauces.

coriander Seed is harvested in the fall. Leaves are gathered at maturity. Ground seed is used in Eastern European cookery, kebabs, stewed fruit, meat loaf, and sausage. Leaves are used in soups and stews.

cumin Seed is gathered at maturity. Dried cumin, an especially hot spice, is used in Mexican and Indian cookery, meatballs, meat loaf, pilaf, any deviled food.

dill Flowers are gathered when they have begun to bloom and still have a few unopened buds. Leaves and stems may be harvested throughout the summer. Seeds may be harvested in fall. Flowers are used in pickles and dilled vegetable dishes; leaves and stems are used in potato, tomato, and some fish dishes. Seeds are used in pickles, coleslaw, creamed dishes, and sauces.

horseradish Roots may be harvested throughout the year. They should be dried whole. They are soaked in vinegar and grated to make salad dressing and meat sauce.

lemon balm Leaves are harvested throughout the year. They are used in tea, as a base for an herbal tea, in fruit salads, and with poultry.

lovage Leaves are gathered throughout the summer. Seeds are harvested at maturity. Leaves and seeds are used to season soups and stews.

marjoram Leaves are gathered when mature. Used to flavor eggs, fish, meat, and chicken dishes.

mint Leaves are gathered throughout summer. Dried leaves are used to make soothing herbal tea.

mustard Seed is gathered in fall. Ground dried seed is used in vinaigrette dressing and rubbed over ham or pork roasts prior to cooking.

oregano Flowers and leaves are harvested just as flowers open and are dried together. Crushed fine, they are added to tomato sauces, stews, and some meat dishes. This is a basic herb of Mediterranean cooking.

parsley Leaves are harvested as they mature. Although dried parsley is not as delicious as when fresh, it is an excellent source of vitamin C. Dried parsley is used in *bouquets garnis*, *fines herbes*, soups, stews, fish, and vegetable dishes.

rose Petals are gathered the day they bloom. Trim off white, tough base and dry on trays. Used to flavor jams, in baked fruit dishes, and to make an herbal tea.

rosemary Leaves are harvested when mature. Used to flavor lamb, chicken, soups, and vegetable dishes.

sage Harvest when leaves are fairly young. Dried ground leaves are used in stuffing for poultry, soups, chicken stocks, and on roast poultry.

sassafras Leaves, roots, flowers, and bark of this tree can be used to make tea. The leaves should be harvested when mature; the root, any time; the flowers, when they bloom. The leaves are used to season soups and vegetable dishes. The ground leaves, called filé, are a thickening and seasoning agent used in gumbo and other Creole dishes.

savory Leaves should be harvested when mature: in the spring for winter savory and in the fall for summer savory. Dried leaves are used in green bean dishes and with poultry.

sorrel Leaves are harvested when mature. Used in sorrel soup, sauces, and egg dishes.

tarragon Harvest leaves when they are young and fairly small. Dried tarragon is used in sauces, fish and shellfish dishes, soups, stuffings, with green vegetables and carrots. *Note:* Only fresh tarragon is used to make vinegar.

thyme Leaves should be harvested as they mature. Flowers should be picked as soon as they bloom. Leaves are used in meat dishes, stews, chowders, game dishes. Flowers, dried and whole, make an herbal tea.

woodruff Harvest leaves when they mature and flowers when they bloom. Used to make an herbal tea.

Recipes Using Dried Herbs

BAKING POWDER BISCUITS WITH DRIED HERBS

These easy-to-prepare biscuits, which adorned the tables of most Americans until the advent of prepackaged dough mixes, are excellent with the addition of dried herbs.

2 tsp. dried, crumbled thyme, dill, marjoram, or a combination of these
¾ cup milk

2 cups sifted flour
1 Tbs. baking powder
½ tsp. salt
¼ cup shortening

Add dried herbs to milk. Sift flour, baking powder, and salt together in a mixing bowl. Add shortening and cut in very thoroughly. Slowly add herb-flavored milk, stirring until mixture becomes a soft dough that pulls away from the sides of the bowl. Use slightly more or less milk if necessary. Turn out on a floured surface and knead a few times. Roll out to a thickness of about ¼ inch and cut into 2-inch-wide circles. Place on a buttered baking sheet and bake in a 450°F. oven for 12 to 15 minutes. Serves 4-6.

CARAWAY CHEESE SPREAD

3 cups grated Swiss cheese
½ tsp. dry mustard
3 Tbs. dried caraway seeds

2 tsp. Dijon mustard
¼ to ½ cup mayonnaise

Mix the cheese, dry mustard, and caraway seeds together. Stir in the Dijon mustard. Add just enough mayonnaise to bind the mixture together. Roll into a ball, wrap in waxed paper, and chill thoroughly before serving on bread or crackers.

CREAM CHEESE AND ONION DIP

1 8-oz. pkg. cream cheese, softened at room temperature
½ cup sour cream

3 Tbs. dried onion
½ tsp. dried thyme
1 pkg. bouillon powder

Mix cream cheese and sour cream. Stir in the onion, thyme, and 1 tsp. bouillon powder. Refrigerate for several hours. Serve with pretzels or potato chips.

SOUR CREAM DIP WITH HERBS

3 Tbs. dried chives   *1 garlic clove, minced*
3 Tbs. dried dill   *1 pint sour cream*

Mix dried chives, dill, and garlic into sour cream. Refrigerate for 1 hour before serving. Serve with crisp raw vegetables, such as carrot sticks, string beans, and cauliflower flowerets.

4

FRUITS AND VEGETABLES

FRUITS AND VEGETABLES

Among the many methods of preserving fruits and vegetables, drying is undoubtedly the easiest and the least expensive. In fact, if you are really determined, initial drying experiments can be set up using materials already available in most homes.

Drying fruits and vegetables is one of the more creative of home food projects. There are few hard and fast rules such as those found in canning and freezing. Much depends upon your urge to experiment and make adaptations to find the methods that suit you best. Because climate and the quality of food vary with each drying session, no two sessions are exactly alike.

Drying methods for fruits and vegetables have changed little over several thousand years. In the Near East, grapes, dates, and figs were frequently buried in hot desert sands until dried. In most Western cultures, drying was done in the sun or shade, outdoors. White settlers in North America learned how to dry food from various groups of Indians, who dried excess berries, fruits, and vegetables, such as corn and beans, to tide them over during the winter months. In some Indian cultures, dried and ground corn was a staple, used much as cooks today use flour. Until the turn of the century, almost every farmhouse cook—and many city and town cooks—put away some dried fruits and vegetables.

This changed as improvements in canning techniques made canning safe for home canners and lessened the interest in drying food. The Depression, World War I, and World War II, however, brought on a wave of renewed interest in drying. People once again found themselves depending upon their own gardens as sources of food. During World War II, sugar and canning supplies were in short supply, so drying was a natural alternative to canning.

Juicy fruits and vegetables must be dried as quickly as possible, as they are excellent breeding grounds for microorganisms, particularly when the foods have been peeled or cut. Four elements cause spoilage in food, and they all are affected by heat: enzymes, molds, yeasts, and bacteria. Their growth and their destruction depend upon the proper application of heat. A food spoils when it is held too long at a temperature that is not high enough to stop or retard bacterial growth. On the other hand, proper drying at the right temperature and done quickly enough retards the growth of dangerous microorganisms.

The sun-drying of vegetables is less successful than that of fruits in most regions of the United States and Canada. Vegetables, which are generally low in acid, also must have more moisture removed during drying than fruits require. Like all low-acid foods, they are more likely to spoil when dried for long periods in the sun. The following vegetables are low in acid content:

okra	*snap beans*
peppers	*greens*
pumpkins	*potatoes*
squash	*asparagus*
carrots	*cauliflower*
cabbage	*lima beans*
turnips	*peas*
beets	*corn*

High-acid foods that are more adaptable to sun-drying include:

green gooseberries	*black and red raspberries*
cranberries	*peaches*
plums	*cherries*
apricots	*pears*
apples	

Success is more readily assured if you begin by drying some of the easier foods. Fruits that are easy to dry include apricots, apples, cherries, dates, figs, and plums. Easy-to-dry vegetables include asparagus, broccoli, carrots, celery, string beans, and green peppers.

Getting Down to Basics

The process of drying fruits and vegetables is a fairly simple one. Basically it involves the following steps: (1) selecting the fruit or vegetable to be dried; (2) preparing the food by cleaning, peeling, pitting, slicing, or whatever is required; (3) pretreatments, which are described in detail later in this chapter; (4) the actual drying process (and again, there are usually several methods to choose from, depending upon climate, space, and food being dried); and (5) posttreatments, which include conditioning, pasteurizing in some cases, packaging, and storing the food properly.

Fruits and vegetables can often be successfully dried in the sun. Solar dryers are also popular and can be used in any area throughout the United States and Canada where there are high temperatures and long hours of daylight. Drying indoors is another alternative and can be done easily in an oven or dryer. In fact, indoor drying in this fashion is the most easily learned technique and frequently is best for persons with limited amounts of space. Some vegetables can be room-dried, just as herbs are. This was a popular drying technique in American pioneer days, but because it produces a tougher product than other kinds of drying, it has fallen into disuse. In fact, it should probably be thought of mostly as a decorative touch for the kitchen. Shade-drying outdoors is another possible method. Obviously, the technique you choose should be based on what is most convenient for you and on the type of food you plan to dry.

Detailed descriptions of how to build and use various kinds of drying equipment were discussed in Chapter 2.

Choosing Fruits and Vegetables for Drying

Although hints on choosing perfect fruits and vegetables are listed under individual foods in the next chapter, there are a few general guidelines for selecting high-quality fruit and vegetables for drying.

Buy directly from an orchard or a produce farmer whenever possible. That way, you will know exactly how fresh and mature the food is. Fruits and vegetables suitable for drying should be fully mature (so that the sugar content is at its peak), unbruised, and evenly colored. If you cannot buy fruits and vegetables from a farmer or some other wholesaler, take your chances with the produce and fruit counter in a large supermarket. Drying results may not be as predictable, since you won't know as much about the quality and freshness of food, but with a little practice, such foods can be easily dried.

The amount of food you purchase depends on how you plan to dry it. An oven will hold 4 to 6 pounds of fruit or vegetables; some dryers suitable for counter top use in homes hold as much as 18 pounds. The capacity varies greatly, of course, with individual dryers and the number of trays used at one time. The chart on the next page shows the yields of dried fruits and vegetables per 25 pounds of fresh food.

Unfortunately, buying fruits and vegetables in large quantities often means they cannot be inspected carefully beforehand. After bringing the food home, sort through it, discarding any pieces that are soft, moldy, or obviously spoiled. If fruits or vegetables have only small bruised or discolored areas, cut these out before pretreating and drying. One piece of blemished food can turn a whole tray bad. Arrange the foods according to size. A batch of fruit to be dried at one time should be composed of pieces that are approximately the same size. Dry only one kind of food in any one batch.

Yields of Dried Fruit per 25 lbs. of Fresh Fruit

apples	3–4 lbs.
apricots	4–4½
blackberries	4–5
cherries	4–6
figs	4½–5½
loganberries	4⅛–5½
peaches	3–4
pears	4¼–5¼
prunes	7¼–8
raspberries	4–5½

Yields of Dried Vegetables per 25 lbs. of Fresh Vegetables

beans	2½–4 lbs.
beets	3¼–4
cabbage	2–2⅛
carrots	2¼–3
celery	2–2⅛
corn, sweet	6¼–8
okra	2¼–2½
onions	2¼–2¾
peas	5¼–6
potatoes, white	5½–8
pumpkin	1¼–2
spinach	2–2¼
squash	1½–2¼
tomatoes	1½–2⅛
turnips	1½–2

Peeling and Checking

The first question that arises when peeling is discussed is whether or not fruits and vegetables should be peeled before they are dried. This is primarily a matter of personal taste. Most commercially dried fruits have been peeled, but it is hardly necessary. Obviously, fruits such as grapes, currants, and cherries are never peeled. Most other fruits can be peeled with a parer or paring knife. Fruits such as peaches, apricots, pears, and nectarines can be peeled more easily if they are first dipped in boiling water for 30 to 50 seconds and rinsed immediately in cold water. After this simple procedure, a few quick slices with the paring knife should be enough to remove the outer layer.

Some fruits, such as cherries, prunes, and some kinds of grapes, have a natural waxy outer coating. Making small breaks in the coating prior to drying speeds the drying process. A quick dip in boiling water—no longer than 45 seconds—serves to break that outer coating. This procedure is called checking.

Old cookbooks and kitchen manuals often suggest using a lye bath to peel or check fruits and vegetables. This method is not recommended today for several reasons. First, household lye, also known as caustic soda, is dangerous to work with and can burn skin surfaces if improperly handled. Second, lye is tricky to work with because it reacts chemically with aluminum, an ingredient in many modern pots and pans. Third, the use of a lye bath destroys many valuable vitamins. But perhaps the strongest reason not to use lye is simply that it isn't necessary—peeling and checking can be accomplished more simply by the procedures just described.

Pretreatments

Anyone who has ever cut up a large quantity of apples or any other highly acidic fruit knows the perils involved. The first pieces have begun to darken before you are halfway through the batch. This happens because the enzyme action that works so well to ripen fruit on the tree or vine speeds up when the fruit is picked. As fruits become too mature, they begin to spoil. In some fruits and almost all vegetables, this spoiling begins when cut pieces of the food come into contact with the air, thus causing the fruit to darken. There are several ways to retard such changes. In many fruits, such pretreatments are optional; in most vegetables, the recommended pretreatments are necessary to retard the growth of bacteria and prevent spoilage during drying. They also preserve the color, check the ripening process, and halt other destructive chemical changes.

Steaming and Blanching

The easiest and most common type of pretreatment is a bath in boiling water or steam for an amount of time that varies with each fruit and vegetable.

Some sources on drying refer to these methods as steam-blanching and water-blanching. To avoid confusion here, steaming will refer to the process to treating food to a steam bath. Blanching will refer to submerging food in boiling water for a brief period of time. It is important to remember that blanching is not cooking.

There are several disadvantages to steaming or blanching; the most obvious is that they add moisture to a food that is meant to be dried. They also cause some deterioration of water-soluble minerals and vitamins. Nutritionists have long debated whether steaming causes less vitamin loss than blanching. There are no certain answers, so choose the method that suits you best.

Blanching can be done in any large pot. The inexpensive canners that show up in stores every summer are ideal. Several gallons of water should be brought to a boil before the food is tossed in. Blanching action is stopped by rinsing the food in cold water for several minutes.

Steaming is done in a smaller pan—a container only large enough to hold a rack for the food. In steaming, the food is never actually submerged in the water.

For $7 or $8, a gadget called a steamer can be purchased in most specialty cookware stores. It does indeed fulfill its purpose, but you can also create your own steamer from a wire basket, a colander, or a large strainer. Simply place it in a pot with about one inch of simmering water. Cover the pot tightly and cook food for the prescribed length of time. Steaming works especially well with apples, apricots, peaches, figs, nectarines, pears, and prunes.

Figure 12. Steamers for blanching fruits and vegetables

The Great Sulfuring Controversy

Sulfuring is the most satisfactory pretreatment for highly acidic fruits (nectarines, peaches, pears) that are to be dried outdoors. Almost all commercial fruits are sulfured. In recent years, however, sulfuring has come under attack from consumer and natural food advocates, who maintain that it is done primarily for cosmetic reasons. While it is true that sulfuring prevents discoloration, it also works to retard souring and helps to prevent insect attacks. In addition, it helps to decrease the loss of vitamins A and C, even though it does speed up somewhat the loss of thiamine (vitamin B_1). Finally, sulfur is itself a mineral—a totally natural product. Food can, of course, be dried outdoors without sulfuring, but it will darken considerably and require extra care during the drying process.

To sulfur, place the cut-up fruit in a small enclosed container in which you are burning some sulfur. The fruit will absorb the fumes of the sulfur for a period of time that varies with the kind and size of fruit. As a general rule, ripe fruit will absorb the sulfur fumes more slowly than immature fruit.

Sulfuring must be done outdoors. Ideally, it should be started in the early evening so the fruit can absorb the fumes overnight.

The first step is to gather the equipment:

Trays used for drying can also be used for sulfuring, provided they are not made of galvanized screening material or aluminum. Sulfur fumes corrode most metal.

Sulfur can be purchased from your local pharmacist. Buy flowers of sulfur, sublimed sulfur, or sulfur candles. Sulfur must be free from impurities in order to burn properly. Do not buy the garden-dusting sulfur, as it is not suitable for working with food.

The sulfur box (see Figure 13) can be any container that is large enough to hold 3 to 4 trays and a small container of sulfur. It can be a cardboard box or a wood packing crate and should be fairly airtight. A small opening or door near the bottom of the box permits air circulation and also allows removal of the sulfur can. In addition, a small flap opening on top of the container is necessary to provide additional ventilation. At least 3 inches of space around the trays is necessary in order for the sulfur fumes to circulate freely.

Bricks or cement blocks can be used to stack the trays inside the sulfur box.

A sulfur pan should be shallow, but deep enough to prevent overflow. A small tin can or foil pan is ideal.

A scale of some sort is needed to weigh the fruit before sulfuring. (If you know how many pounds you have, this piece of equipment is not necessary.)

A teaspoon is needed to measure the sulfur.

Prepare the food for sulfuring by washing, peeling (if desired), and pitting it. Halve or slice it. Place the fruit cavity side up, on trays. Weigh fruit before you arrange it in single layers on the trays. Line the trays with cheesecloth, if desired. During sulfuring, syrup collects in the cavity of fruits. Care should be taken not to spill this syrup while the fruit is being weighed and transferred to the dryer.

The amount of sulfur used depends upon the amount of fruit dried. Use 1½ to 2 teaspoons per pound if sulfuring will take less than 3 hours; use 3 teaspoons per pound if sulfuring takes more than 3 hours. Required times for sulfuring are listed under individual fruits later in this chapter. Sulfur burns best if it is not more than ½ inch deep. Place the sulfur can on the ground where you plan to sulfur and light the sulfur with matches (see Figure 13). Do not leave the matches in the can. Stack the trays in staggered fashion slightly behind the sulfur can and place the sulfur box over everything. Leave the flaps open while the sulfur is burning, which takes about 5 minutes. When the sulfur stops burning, close the flaps and begin to count the sulfuring time.

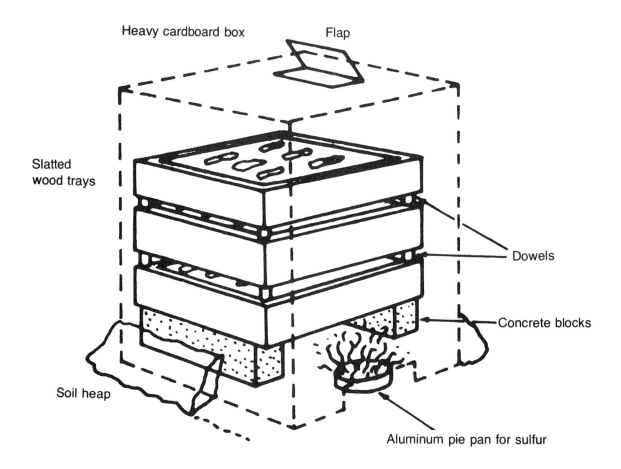

Figure 13. Sulfuring

Syrup Blanching

Candied dried fruit is often used in fruitcake, breads, and desserts. It is made by blanching fruit in a sweet syrup rather than sulfuring it. Fruit that has been blanched this way takes several hours longer to dry, and if it dries outdoors, it must be carefully protected from insects. If possible, it should be dried in an oven or dryer.

Make a blanching syrup by combining 1 cup sugar, ¾ cup corn syrup, and 4 cups water. Simmer 2 cups minced (mixed, if you like) fruit for 10 minutes. Remove from heat and cool in syrup 20 minutes. Drain the fruit and cool before drying.

Sodium Bisulfite Bath

If you live in a place where sulfuring is not possible, the use of a sodium bisulfite solution is a good alternative. It can be obtained from pharmacists or chemical companies. Mix 1½ to 2 teaspoons of sodium bisulfite in each gallon of water used. Fruits should be soaked for the following times in this solution:

apricots	soak 10-15 min.
nectarines	steam 5 min., cool to room temperature, soak 15-20 min.
peaches	steam 5 min., cool to room temperature, soak 15-20 min.
pears	soak 20-25 min.

Ascorbic Acid Bath

Ascorbic acid baths are a less satisfactory way to prevent discoloration than either sulfuring or a bath in sodium bisulfite solution, but they can be used if necessary. The primary disadvantage of an ascorbic acid bath, and also of the sodium bisulfite solution, is the addition of moisture to a food that will be dried. In the case of the ascorbic acid bath, there is also some loss of water-soluble vitamins, minerals, and sugars. Still, it works well enough to be worth the effort if you have no alternative. Buy ascorbic acid from a pharmacist. Vitamin C in powdered form also works well, if you can find it at a pharmacy, or make special arrangements to purchase it through a chemical company.

Dissolve 2 teaspoons ascorbic acid or vitamin C in each cup of cold water and drop halved and pitted fruit into the bath. Work quickly so the cut-up fruit does not stay in the bath very long. As soon as enough food is prepared to fill a tray, do so. Drain and pat food dry and place in a single layer on the tray. Begin drying at once.

Predrying Suggestions for Fruits

FRUIT	SIZE OF PIECES	TREATMENT	TIME REQUIRED
apples	thin slices	sulfur	20-30 min.
		steam	5-7 min.
	quarters	sulfur	45-60 min.
apricots	halved and pitted	sulfur	90-180 min.
		steam	5-7 min.
berries (blackberries dewberries, loganberries, raspberries)	whole	none	
cherries	whole	checking required; pit fruit and drain for 1 hour; reserve juice for other uses	
figs	halved and pitted	steam or blanch (cut figs only)	20 min.
		sulfur	see individual instructions
nectarines	halved and pitted	sulfur	90-180 min.
		steam	5-7 min.
peaches	pared halves	sulfur*	20-30 min.
	unpared halves	sulfur	60-180 min.
		steam	15-20 min.
	quartered or in thinner slices	sulfur	2-3 hrs.
		steam	5-8 min.
pears	quartered or in thinner slices	sulfur	2-3 hrs.
prunes	whole, dark-fleshed	check 30-45 sec.	
	whole, light-fleshed	check and sulfur	20-30 min.
	halved	same treatment as peaches; take care to save juice in cavity	
		steam	15 min.
	slices	steam	5-8 min.

---

*Time varies greatly: soft-fleshed varieties may need only ½ hr.; firm-fleshed varieties, up to 2 hrs.

Drying Methods

There are several ways to dry fruits and vegetables. The method you choose should be both convenient and suited to the facilities available to you.

Sun-Drying

In order for sun-drying to be truly effective, the climate must be nearly perfect—quite dry with consistently high temperatures. A strong breeze is also helpful. Since fruits and vegetables should be dried when they are at the peak of maturity, outdoor drying is dependent on several uninterrupted sunny days with no precipitation. Sound impossible? It pretty much is in all except a few areas of the United States. Southern California has a perfect climate for sun-drying, and most of the commercially dried fruits come from that region. Areas of the Southwest —Arizona and New Mexico—also have trustworthy climates for sun-drying. You can always try sun-drying in other areas, but the results cannot be guaranteed.

Foods dried outside also need a minimum of pollution, not an easy thing to come by these days. It is best not to assume that all fresh air—even in the country —is safe. Call a county agent to double-check the local pollution factors before you begin drying. Finally, never dry food near a busy highway or a dusty country road. The former has lead pollution, and the latter is just plain dirty.

Foods dried outdoors are placed on trays and covered with a light protective material. Since most fruit becomes sticky as it dries, it helps to place a layer of cheesecloth on the bottom of the tray, or alternately, to coat the tray bottoms with mineral oil, which will prevent stickiness or the flavor of the fruit from seeping into the wood. It also makes cleaning the trays an easier task. Cheesecloth on top of the trays prevents insects from getting to the fruit and will help to keep off dust. The ideal place to dry fruits and vegetables outdoors is the slanted roof of a building, something like the roofs found on chicken coops. Lacking your own chicken coop, you can place the trays in any sunny area off the ground. Use cement or wooden blocks to prop the trays off the ground and separate them. The trays should be about 6 inches off the ground and have 3 to 6 inches of space between them in order for the air to circulate freely.

City dwellers will find the roofs of apartment buildings excellent for sun-drying, providing the pollution level is low enough to make it safe.

An electric fan can be used to speed sun-drying. Turn it directly on the trays, taking care to shift the fan or the trays from side to side so all the fruit is equally exposed to the breeze.

If you want to try your hand at sun-drying without going to the trouble and expense of building trays and purchasing other equipment, simply stretch a sheet or several layers of cheesecloth on the ground or on a picnic table. A sheet is heavy and doesn't provide good air circulation, but it will do in a pinch.

Regardless of the materials used, place the food in the sun in the early morning. Stir or shake the food several times a day to encourage even drying on all sides.

Around sundown, move the food inside or put it in a sheltered area for the night. Wait about 30 minutes for it to cool down and then cover it with a sheet. If the food is covered right away, moisture could form.

Return the food to the sun early the next morning. By the end of the second day, begin to check individual pieces for dryness. As pieces appear dry and meet the test for dryness (see section on this later in chapter), remove them from the trays and place them in a container to which you can later add the rest of the dried pieces.

Shade-Drying

Drying food in the shade is slightly trickier than other kinds of drying; its success depends on the food and the weather conditions.

Shade-drying requires trays, cheesecloth, and an electric fan that can be used to encourage air circulation. The fan also works to speed the drying process, which is especially important in drying low-acid foods that spoil quickly. Food is most easily shade-dried in a dry climate with a strong breeze and consistently high temperatures.

Cut the food into smaller pieces than you would for sun-drying. Then arrange it in shallow layers on the trays. Although most of shade-drying is done in a shady spot, it is wise to start the food in the sun. Stack the trays in a sunny spot; they should be at least 6 inches apart. Shift the trays around 5 or 6 times a day to insure even drying; also shake the trays frequently to shift the food. After 1 to 1½ days in the sun, or when the food begins to look dry, remove the trays to a shady area and drape them with cheesecloth. Turn the electric fan on them. Take care to shift the position of the trays even more frequently—every 45 minutes or so—so the breeze hits all the food for equal amounts of time. Begin to test the food for dryness after the first full day of shade-drying. Remove individual pieces of food as they are done. When the drying is completed, condition the food and pasteurize it. This method of shade-drying, using an electric fan for ventilation, also works indoors.

Solar Drying

If you do not want to invest a lot of money and have only a limited amount of space for outdoor drying, consider making and using a solar dryer. A solar dryer can be used on a rooftop or a patio. Complete instructions for building a solar dryer can be found in Chapter 2.

The operation of a solar dryer, as environmentalists are well aware, costs nothing. It is also slightly faster than sun-drying in open air. While it only dries small amounts of food at a time, this is not necessarily a drawback for an urban dweller, who may have only limited amounts of storage space.

Cut the food into fairly small pieces, and pretreat it as required for foods that will be dried in the sun. Spread the food in a single layer in the dryer, making sure to open all vents to allow circulation. Check the trays every 45 to 60 minutes—be sure that all the pieces of food are exposed to the sun's rays. The dryer should be placed in direct sunlight at all times, so do not forget to move it around as the sun shifts during the day. Around 3 or 4 P.M., when the sun's rays are no longer strong enough to dry food, close the dryer's vents, let the food cool down, and place a sheet or other protective material over the dryer. Begin the drying process again the next morning. Food dried in a solar dryer generally takes about 2–3 days, or slightly longer than sun-drying. Remove individual pieces as they meet the test for dryness. Condition the food and pasteurize it. Store carefully.

Room-Drying

Room-drying is not a particularly reliable method of drying food. In fact, you should experiment with a small quantity of food before you make a major investment in bushel quantities of fruit or vegetables for room-drying.

Frontier kitchens were often adorned with strings of drying apples, squash, or pumpkins. When the apples were cooked, they were called "leather britches," which should give some clue as to the chewability of foods dried this way. Wood-burning stoves provided the right proportion of warm, dry air and low humidity. If you can meet these climatic conditions, it is fun and decorative to try to room-dry some foods.

Room-drying at best is tricky, however, and if you are just gaining experience in home drying, you would do better to start with either of the two indoor drying methods that follow.

Drying in an Oven or Homemade Dryer

Indoor drying is better than sun- or shade-drying or, for that matter, any type of outdoor drying, simply because it requires less time. A shorter drying time means better color preservation, flavor, and overall quality. Fewer nutrients are lost through indoor drying because the process is faster. Another advantage is that drying can go on continuously since it is not limited by the amount of available sunlight, but only by the willingness of people to watch over the drying process.

Trays that are used for other kinds of drying—wood or nongalvanized metal—work well in ovens and dryers. Cookie sheets and other baking dishes are not adequate because they prevent the free flow of air around the food.

To dry in an oven, remove the oven shelves and replace them with trays spaced about 6 inches apart. Avoid using cheesecloth as it catches fire too easily. The only other equipment needed is a good thermometer—a meat thermometer will do quite nicely.

Either a gas or an electric oven can be used. The temperature is regulated by the oven controls and by opening the door of the oven. Use a thermometer to check the temperature. The door should be slightly more ajar in a gas oven than in an electric one; 1 to 8 inches is enough, depending upon the drying temperature required. In an electric oven, only the lower heating coil (the one used for baking) should be used. If you cannot regulate this with the controls of the oven, disconnect the broiler or top heating coil.

The tray closest to the heating coil will dry fastest—it should be at least 10 inches away from the heating apparatus. The top tray dries next. Trays should be shifted frequently throughout the drying process.

Although most ovens hold a maximum of 4 to 6 pounds of food, food does not dry evenly if it is crowded. It is better to dry fewer trays of food, if necessary, to maintain enough room for air to circulate around the trays.

Begin by heating the oven to 140°F. If you cannot hold the temperature here, remove the trays even farther from the heat—if necessary, drying one less tray.

Oven-drying takes a little more attention than other methods, because the food should be shifted every 30 minutes or so and the temperature must be regulated very carefully to prevent scorching. Watch the oven carefully the first few times you use it to see how quickly and how evenly it dries, and accommodate your technique accordingly.

Food preparation for oven-drying (or any indoor drying, for that matter) is about the same as for outdoor drying. You should omit sulfuring, but you can blanch or steam the food according to the directions listed with individual fruits and vegetables on page 60.

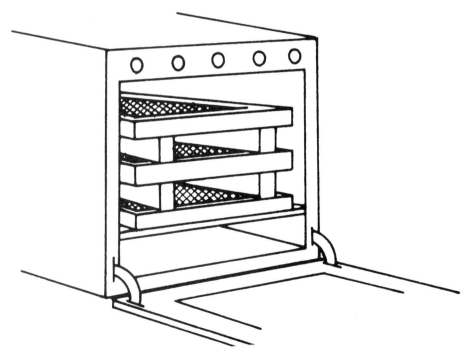

Figure 14. Oven-drying

Using a Homemade Dryer

A dryer is perhaps the most efficient way of drying fruits and vegetables. It generally holds a larger amount of food than an oven and small dryers work perfectly well in a little kitchen, so they are an easy way for apartment dwellers to dry food. Like an oven, however, a dryer demands a certain amount of watching.

Set the dryer at 140°F. and see how easy or difficult it is to hold it there. If there are problems, you should watch the temperature carefully to make sure that the food is not cooking—as opposed to drying. If, at any time, the food appears to be drying out too quickly (it dries fastest during the first few hours), reduce the temperature at once by 10 to 15 degrees. Later, when most of the moisture has left the food, the rate of drying slows down and the chances of scorching the food speed up. The last few hours of drying time—regardless of the method—require careful attention.

Arrange the food in single layers on trays, shifting it every 30 to 45 minutes to insure even drying. As food shrinks and dries, move it to trays closer to the source of heat. The food can now be several layers deep on each tray. If new food is added at this point—and there is no reason not to add new trays of food as space permits—it should be placed in single layers on trays that are positioned at the topmost part of the dryer.

Testing Food for Dryness

Begin testing as soon as the food appears dry and shriveled—about 2/3 of the way through the drying process. It still needs more drying time, obviously, but testing will help develop your ability to spot perfectly dried food. To test, remove one or more pieces of food and allow them to cool to room temperature. This is necessary because food, when hot, seems moister and more pliant than it actually is. Tests for dryness are listed under individual fruits and vegetables in the next chapter.

It is helpful to keep a notebook of your experiments in drying foods. Record drying times, weather conditions, and the hours when food is dried, plus any special techniques that work well for you. You will then be able to go back to these techniques, altering or developing them as you become more proficient.

Conditioning

Conditioning is necessary because not all foods lose the same amount of moisture as they dry. Some portions of a piece of food may become overly dry; others, not quite dry enough. During conditioning, the moister portions of food give up their moisture to the dryer portions. A state of equilibrium is achieved so that foods will neither absorb moisture nor give it off. This prevents bacterial growth when the food is stored. Conditioning is the finishing touch that will insure longer life for dried foods.

Conditioning should be done in a dark place. Around the turn of the century, many cooks conditioned food on the scrupulously clean floor of a pantry. Today, food is frequently conditioned in bins, plastic containers, a burlap bag—any clean container. While food is conditioning, stir it frequently, once or twice a day. Conditioning takes about 10 to 12 days. Check the moisture condition of food every day; perceptible changes will occur. When these day-to-day changes seem to have stopped, the food is properly conditioned.

All the food from one drying session should be conditioned at the same time. In fact, several days' worth of drying and several kinds of foods (all fruits or all vegetables) can be conditioned together.

Pasteurization

Pasteurization is necessary for all fruits and vegetables dried outdoors. It is also necessary for foods that are dried indoors at temperatures of less than 150°F. Even if a food has been dried at 150° but has not been held at that temperature for more than 30 minutes, it should be pasteurized. Food should be pasteurized after conditioning.

To pasteurize food, spread it on trays (or cookie sheets or other baking pans) and heat in a 150°F. oven for 30 minutes. Cool to room temperature before storing.

Storage

Dried fruit or vegetables should be stored in a sealed container and placed in a dry, warm place. Whether or not the container should be airtight is a debatable subject. In warm climates, water often condenses from the air in the containers and causes spoilage on the surface of the dried food. Airtight containers that are opened frequently are probably safe for small quantities of food. In humid climates, dried foods are best preserved by refrigeration.

If fruit becomes moist during storage—and you should check for this about once a month—heat it in a 300°F. oven for about 30 minutes and then return it to its containers.

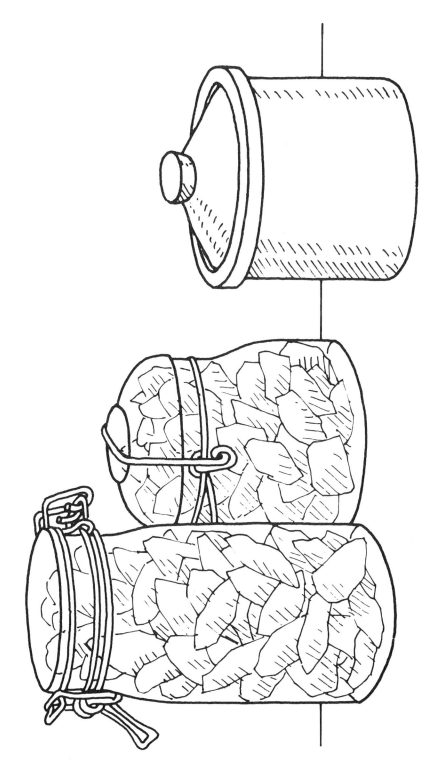

Figure 15. Containers for storing fruits and vegetables

Unpleasant as it may be to discuss, bugs and worms do occasionally appear in dried foods, particularly in food that has been dried outdoors. Pests lay their eggs when the fruit is drying, and these eggs may not hatch for several months. Pasteurization and proper drying minimize pests, but should they appear, there is an easy solution to the problem. Heat the food to 300°F. for about 30 minutes. This will destroy the invaders and sterilize the fruit, so you need have no worries about eating it.

Properly conditioned and stored, food keeps for up to one year.

5

TECHNIQUES FOR DRYING

TECHNIQUES FOR DRYING

This chapter presents suggestions on the varieties of fruits and vegetables that dry well, their average drying times, and individual tests for dryness.

Apples

Selection. Use slightly tart "eating apples" as opposed to "cooking apples." Rome Beauty, Northern Spy, Jonathan, Baldwin, Winesap, Gravenstein, Golden Delicious, and Red Delicious work well. Choose mature fruit that is evenly colored and unbruised.

Preparation. Wash and peel apples. Optionally drop fruit in an ascorbic acid bath to prevent browning. Blanch or sulfur (see table in Chapter 4 for length of time). Arrange in tray no more than 2 inches deep.

Indoors. Start in dryer at 130°F.; increase to 165°F. after about 1 hour. When most of the moisture has left the fruit, lower the temperature to 145°F. and regulate it between that temperature and 130°F. as needed to prevent scorching. Stir and turn fruit on trays frequently as it finishes drying. Apples should dry in 6 hours or less under controlled heat.

Outdoors. Dry in the sun for 1 to 3 days. Sulfuring is required. Place the fruit one layer deep on trays and cover with cheesecloth or other light fabric during drying. Condition and pasteurize.

Test for dryness. Apples are properly dried when they take on a leathery, suedelike quality and feel springy to the touch. When squeezed, they should leave no feeling of moisture in the hand and should separate when the fist is released. Apples are overdried when they become crisp and hard; such pieces should be discarded.

Apricots

Selection. Any variety may be dried, but Blenheim, Royal, and Tilton work especially well. Ideally, apricots should be picked before they fall from the tree, when they are fully matured, with full color but still quite firm. Avoid fruit with bruises and discolored spots.

Preparation. Apricots usually are not peeled. Wash and dry the fruit. Halve small apricots and cut larger ones into quarters or slices. Remove pits. An ascorbic acid bath may be used to prevent browning during preparation. Remove from bath and place in trays as soon as you have enough to fill one tray. Steam or sulfur. (If sulfured, do not use ascorbic acid bath.) After blanching or sulfuring, arrange the fruit in single layers in trays, taking care not to spill any syrup that has gathered in the cavities during sulfuring.

Indoors. Start drying at 130°F. After 1 hour, raise the temperature to 150°F. As the apricots show signs of dryness, lower the temperature to 130°–140°F. to finish drying. Watch carefully for scorching toward the end of the drying and remove any apricots that are done. Stir and shake the trays frequently as fruit dries. Apricots dry in about 14 hours in a dryer or oven.

Outdoors. Apricots may be sun-dried or solar dried. Follow the same preparation procedures as for indoor drying.

Test for dryness. Cut the center of the fruit. If there is no moisture, it is properly dried. Grab a small handful of the dried apricots and squeeze them in your hand —they should leave no moisture on your palm and should separate easily when your fist is released. Their texture should be leathery and pliant.

Berries

Selection. Firm varieties—blackberries, loganberries—are easily dried. Red raspberries and strawberries are best left to commercial dryers; they are too moist to dry well even under the best of home drying conditions. The best berries are those you pick yourself, but if you know your greengrocer and can rely on his assurances that the fruit is fresh, you can use store-bought fruit. Fresh-picked fruit should be harvested in the early morning before it has been heated by the sun.

Preparation. Sort through the fruit to remove any bruised or moldy pieces. Leave berries whole and do not attempt to remove skins. Waxy varieties of berries should be checked for 15 to 30 seconds, depending upon their size.

Indoors. Spread berries one or two deep on trays that have been covered with cheesecloth. Start drying at 120°F. After one hour, increase temperature to 130°F. When berries have shriveled and are beginning to look dried, raise temperature to 140°F. and keep it there for 2 hours. Reduce heat to prevent scorching as berries finish drying. Drying time is 4 hours.

Outdoors. Spread berries on trays and dry 1 to 2 days in sun or solar dryer. Condition and pasteurize. After berries have dried, remove any berries that may have stuck together.

Test for dryness. Berries are properly dried when they are hard and rattle when shaken in the tray. They should not release any moisture when crushed between the fingers.

Cherries

Selection. Both sweet and sour varieties of cherries are easy to dry. Choose firm, unbruised fruit.

Preparation. Clean and pit the cherries or leave whole. Check 15 to 30 seconds, depending upon size of fruit. Place in single layers in trays covered with cheesecloth.

Indoors. Dry at 120°F. for 1 to 2 hours. Raise the temperature to 145°F. for a few hours and finish by gradually reducing the heat. Check carefully for scorching. Drying can take up to 6 hours.

Outdoors. Pit cherries to speed drying. Arrange in single layers on trays covered with cheesecloth. Cherries will sun-dry in 1 to 2 days. Condition. Pasteurize.

Dates

Selection. Choose firm, translucent fruit. Discard any fruit that has fungus, bruises, a sour smell, or is soft. Halve dates if they are large. Wipe each fruit with a damp towel. No pretreatment is needed, but only one variety of date should be dried in a batch.

Indoors. The best dried dates are produced in your own oven. Preheat the oven to 225°F. Turn off the heat and place the fruit, either on drying trays or on a cookie sheet, in the oven. Leave in the oven until the fruit has cooled. Repeat the process the next day. Two days are needed to dry by this method.

Outdoors. Dates will dry in a solar dryer in 2 to 8 days; they will dry in the sun on trays in approximately the same amount of time. Condition. Pasteurize at 150°F. for 30 minutes.

Test for dryness. Dry dates are leathery, pliant, and slightly sticky.

Figs

Selection. Choose fully ripe fruit, or the sugar content will be too low and the fruit will sour as it dries. The best varieties for drying are Adriatic, Mission, and Kadota.

Preparation. Wipe figs with a damp cloth to clean. If the figs are small, leave whole; otherwise, cut in half. If the figs will be dried outdoors, check them for 30 to 45 minutes before sulfuring for 1 to 2 hours. Note that only the light varieties can be sulfured; black figs will mottle. Cut figs that are dried indoors should be steam-blanched for 20 minutes.

Indoors. Start oven or dryer at 115°F. and raise to 140°–145°F. after 1 hour, or when fruit has noticeably begun to shrivel. Reduce temperature to 130°F. or lower for the last hour and watch carefully to avoid scorching. Stir fruit frequently to prevent sticking. Five hours is required to dry halves; a few more for whole figs.

Outdoors. Arrange the figs in a single layer on cheesecloth-covered trays. Turn frequently. Several days will be needed to dry thoroughly. *Note:* Some friends who have had considerable success in drying fruits on their farm in southern Indiana report that they dip dry figs into a boiling brine solution composed of nearly 1 pound of salt in 3 gallons of water for 2 minutes. They have found that this final treatment softens the skins and makes them glossy. It also sterilizes the figs.

Grapes

Selection. Any seedless grape may be dried, but experienced driers have reported especially good luck with Thompsons and Muscadines. The quality and flavor of grapes may vary with the type of grape dried; Muscadines, for example, are sweet and hold their flavor well.

Preparation. Wash and dry the fruit and remove any stems and leaf bits. No treatment is required unless the skins are waxy, in which case the fruit should be checked for 15 to 20 seconds.

Indoors. Start the oven at 120°F. and gradually raise to 150°F., taking care to stir the grapes frequently. Indoor drying takes approximately 8 hours.

Outdoors. Grapes are easily dried in the sun on trays lined with cheesecloth. Two or 3 days are required.

Test for dryness. Dry until pliable and leathery—in short, until the grapes have become raisins.

Nectarines

Selection. If possible, nectarines should be picked before the fruit is ripe enough to drop from the tree. If they must be purchased from a greengrocer, select the firmest, most evenly colored fruit available.

Preparation. If you want to peel the fruit, dip it for 15 to 20 seconds in a boiling water bath. The fruit, depending upon its size, should be cut into halves or quarters. Fruit dried inside should be dropped in an ascorbic acid bath until it is dried. If drying outdoors, steam before sulfuring: 15 to 18 minutes for whole nectarines and 5 minutes for halves or smaller pieces. Be sure to transfer sulfured fruit to trays without spilling any of the syrup that has collected in their cavities during sulfuring.

Indoors. Single layers are necessary because of the collected syrup. Start at 125°F. and hold this temperature until the syrup has evaporated or is absorbed. Turn fruit over. Go to 155°F. and hold at that temperature until fruit is nearly dry. When fruit begins to show signs of a dry appearance, reduce the temperature to 130°F. and maintain that temperature until perfectly dried. Stir and shake the fruit frequently after it has begun to dry to prevent scorching. Drying time is 15 hours for halves and 6 hours for slices.

Outdoors. Dry in single layers on cheesecloth-covered trays in the sun or in a solar dryer for 3 to 4 days. Condition and pasteurize.

Peaches

Selection. Any good "eating" as opposed to a "baking" peach is suitable for drying—freestone and yellow-fleshed varieties produce the best results.

Preparation. Commercially dried peaches are never peeled, but if you want to peel peaches, dip them first into a boiling water bath for 15 to 20 seconds. Rinse in cold water. Skins will slip off easily. Cut in half and pit; if peaches are very large, drying will be speeded if they are cut into smaller pieces. Scrape away the reddish pulp, as it turns dark during the drying process. If drying inside, treat with an ascorbic acid bath to prevent browning. Steam halves for 15 to 20 minutes, slices for 5 to 8 minutes, before placing in dryer or oven. Fruit being prepared for outdoor drying should be sulfured. The sulfuring time can be cut in half if the fruit is steam-blanched first. Transfer to drying trays without spilling any of the syrup that accumulated during sulfuring.

Indoors. Start at 125°F. and hold that temperature until syrup has vanished. Turn fruit over and increase temperature to 155°F., holding there for about 7½ hours and watching carefully to be sure fruit does not become overly dry. When fruit is nearly dry, reduce heat to 125°–130°F. and watch carefully to prevent scorching.

Outdoors. Fruit need not be peeled for sun-drying. Dry slices in solar dryer or sun for 3 to 4 days. Halves should be dried for 4 to 5 days or longer. Turn over first time when syrup has vanished from cavity. Turn frequently during drying. Condition. Pasteurize.

Test for dryness. Fruit should be leathery, somewhat pliable, and show no signs of moisture when cut. It should feel slightly rubbery to touch.

Pears

Selection. Bartlett pears are the best variety to dry. Buy ripe, but very firm, unbruised fruit with no soft spots. If picked or purchased very recently, the fruit may be stored in a cool place for 1 week, until ripe enough for drying. If purchased in a grocery store, the pears should be bought at the right stage of ripeness and dried within 24 hours of purchase.

Preparation. Pare, halve, and core. Cut into quarters or thin slices and treat with an ascorbic acid bath to prevent browning. If pears are going to be dried in an oven, steam them for 5 to 20 minutes, depending upon the size of the pieces. If drying outdoors, sulfur. Spread pears on trays in single layers.

Indoors. Start at temperature of 130°F. and gradually increase to 150°F. (usually about one hour into the drying process). Reduce to 130°–140°F. for last hour or so, while drying process is slowing down and almost completed. **Watch carefully** for scorching; shake and stir fruit frequently. Drying time will be about 15 hours for halves and about 6 to 8 hours for slices, depending upon their size.

Outdoors. It is difficult to dry pears outdoors and not recommended for beginners or those with a less than perfect climate.

Test for dryness. Dry until springy and suedelike. Make a small test cut through a piece to be sure there is no moisture inside.

Prune Plums

Selection. Prunes—or rather, plums raised to become prunes—were introduced in California in 1856 by a Frenchman named Pierre Pellier. The Agen plums he introduced are one of the oldest and best varieties for drying, and you should make every attempt to find this kind for home drying.

Preparation. Dry the plums whole if they are small enough. Otherwise, cut them into halves and remove the pit. For indoor drying, steam plum halves for 15 minutes, slices for 5 minutes. In drying outdoors, check the skins of whole fruit. Sulfur for 2 hours for whole fruit, 1 hour for halves or slices. Arrange fruit in a single layer in trays.

Indoors. Start at 130°F. and gradually increase temperature to 150°F. Hold at this temperature for 4 hours, then decrease heat to 130°F. for last few hours of drying. Slices and halves will dry in 6 to 8 hours; whole plums will require 14 hours.

Outdoors. Fruit dries in the sun in 14 to 24 hours. Condition and pasteurize.

Test for dryness. Prunes should be pliant and leathery. A handful should spring apart after squeezing in your fist. Finished prunes should have a chewy texture.

Artichoke, Globe

Selection. Only the tender hearts can be dried. Select young, tightly closed artichokes with few bruises. Save the leaves to steam and eat later; discard the choke.

Preparation. Cut the hearts into ½-inch chunks or strips. Depending upon the size, blanch for 6 to 9 minutes in boiling water. Drain and rinse in cold water. Arrange on trays in single layers.

Indoors. Drying time in a dryer is about 3 to 5 hours; oven-drying time, about 4 to 6 hours. Start at 140°F.; increase to 150°F. and hold for about half of the drying time, or until the vegetable starts to look and test dry. Reduce heat to 130°F. for the last hour or so of drying time. Watch very carefully.

Outdoors. Preparation is the same as for indoor drying. Takes approximately 8 to 10 hours in sun, slightly less in solar drying. Condition and pasteurize.

Test for dryness. Brittle.

Asparagus

Selection. Dry the tips only for best results. Choose tender, tightly closed pieces.

Preparation. Wash thoroughly and cut larger tips into small pieces. Steam for 4 to 5 minutes.

Indoors. Start at about 140°F. and hold this temperature for 1 hour; increase heat to 150°F. and keep until vegetable is almost dry. Finish drying at 130°–135°F.

Drying in a dryer takes about 1 to 4 hours, depending upon the size and age of the asparagus. Drying in an oven takes slightly longer: 3 to 4½ hours.

Outdoors. Allow 8 hours. Process in same way as for indoor drying. Condition and pasteurize.

Test for dryness. Leathery to brittle.

Beans, Green (also called string or snap beans)

Selection. Choose the smallest, tenderest beans you can find, preferably a stringless variety.

Preparation. Wash thoroughly and wipe dry with a towel or paper. Blanch or steam 6 minutes.

Indoors. Start temperature at 120°F. and keep for 1 hour. Increase to 150°F. and keep this temperature for about half the drying time. Reduce to 120°–130°F. when beans are almost dry. Oven-drying takes about 3 to 6 hours; dryer time, 2¼ to 4 hours. If beans have been split, start at 130°F. in order to evaporate surface moisture more rapidly. Beans can be dried in a light, breezy room fairly easily. Blanch as for other kinds of drying. Do not split. String the beans at one end with clean string, keeping the beans about 1 inch apart. Hang in a dry, well-ventilated room or near a warm stove. *Note:* Do not hang over a burner where you will be steaming or boiling foods. The beans should dry within a few days.

Outdoors. Place beans on trays 1 layer deep. They will dry in the sun in about 8 hours.

Beans, Lima (also called shell)

Selection. Permit the beans to reach fuller maturity in their pods than if you were to pick them for immediate eating, but take care to pick them before the pods have dried out.

Preparation. Shell. Steam for 8 minutes. Spread on trays in a single layer.

Indoors. Start at 140°F. and hold for 1 hour. Raise temperature to 160°F. and hold for about half the drying time, or until the beans are showing signs of dryness. Finish for the last hour or so at 130°F. Condition and pasteurize. Drying takes 3 to 6 hours in oven; 2 to 4 hours in dryer.

Outdoors. Sun-drying and solar drying do not work especially well with this type of bean.

Test for dryness. Hard, brittle. Breaks clean when broken.

Beets

Selection. Choose small, tender, unbruised vegetable.

Preparation. Clean and remove all but ½ inch of tops; this will prevent bleeding as beets cook. Cook the beets for 30 to 50 minutes, in just enough boiling water to cover them. Cooking time will depend on size of beets. The cooking takes the place of blanching or steaming. Rinse in cold water and pat dry with towel. Peel and cut into ⅛-inch-thick slices. Spread beets in trays in single layers.

Indoors. Drying takes 2 to 3 hours in dryer; in oven, 3 to 6 hours. Start at 120°F. and increase temperature to 150°F. after the first hour. Keep at 150°F. for about half drying time, or until beets look dried, then drop temperature to 130°F. and watch carefully during last hour of drying.

Outdoors. Beets dried outdoors should be shredded in order to facilitate drying. It takes 8 to 12 hours. Condition and pasteurize.

Test for dryness. Beets will have dry, leathery texture; shreds will be brittle enough to break. Color will change to dark red as beets dry.

Broccoli

Selection. Choose young, tender stalks.

Preparation. Split the stalks into quarters or eighths, depending upon their size. Steam 8 minutes, or until stalks seem tender when pierced with a fork.

Indoors. Start in oven or dryer at 120°F. and increase to 150°F. after first hour. Keep at 150°F. for about ½ of drying time, then reduce to 140°F. when nearly dry. Watch carefully to avoid scorching. Test for dryness. Condition and pasteurize. Broccoli will require about 2½ to 5 hours in a dryer and 3 to 5 hours in an oven.

Outdoors. Spread in single layers in trays. Drying takes 8 to 10 hours in the sun and slightly less time in a solar dryer. Condition and pasteurize.

Test for dryness. Brittle.

Cabbage

Selection. A small quantity of dried cabbage is excellent in soup. Select an unbruised head and wash and remove the outer leaves.

Preparation. Shred coarsely. Steam or blanch for 5 minutes. Drain and pat dry. Spread evenly about ¼ inch deep in trays.

Indoors. Start the oven or dryer at 120°F. Increase to 140°F. after 1 hour, then reduce to 130°F. as soon as cabbage takes on dried appearance. Cabbage has a tendency to mat as it dries, so plan to stir and shake it quite frequently. About 1 to 2 hours is needed for a dryer and 2 to 3 hours for an oven.

Outdoors. Seven to 8 hours drying time is needed in the sun and slightly less in a solar dryer. Condition and pasteurize.

Test for dryness. Ribs are very tough and thin leaves crumble easily.

Carrots

Carrots are so widely available and keep so well in cold or root-cellar storage that there is practically no reason to dry them. Should you have no other means of obtaining carrots, however, here's the technique:

Selection. Choose tender, small carrots.

Preparation. Peel or not according to personal taste; young carrots really don't need it. Cut crosswise in slices or shred. Steam or blanch for about 4 minutes. Rinse in cold water to stop cooking action and pat dry with a towel.

Indoors. Start oven or dryer at 120°F. Increase to 150°F. for about ½ the drying time. Reduce to 135°–140°F. during last stages of drying and check progress carefully. One to 2 hours will be needed in a dryer, 2 to 3 hours in an oven.

Outdoors. Dry for 8 hours in the sun and slightly less in a solar dryer. Use shredded carrots for best and fastest drying results.

Test for dryness. Tough.

Cauliflower

Selection. Choose small, very fresh flowerets.

Preparation. Break or cut into small chunks. Blanch for 3 minutes; drain and rinse in cold water. Pat dry with paper towels. Spread in single layer in drying trays.

Indoors. Start in oven or dryer at 120°F. After one hour, increase heat to 150°F. and hold for about half of drying time. Reduce heat to 125° to 130°F. to finish drying. About 2 to 3 hours is needed in a dryer and 4 to 6 hours in an oven.

Outdoors. Dry on trays in the sun for 8 to 12 hours. Condition and pasteurize.

Test for dryness. Texture should be tough, even brittle.

Corn, Sweet

Sweet corn is one of the few dried vegetables that is far superior to anything that comes in a can.

Selection. Good varieties to look for are Stowells Evergreen, Country Gentlemen, Golden Bantam, or any other table-ready sweet corn. The ears should still be in the milk stage.

Preparation. Husk the corn and use a sharp knife to cut the corn off the cob. (*Note:* The cobs make excellent fuel for smoking meat and fish.) There is no need to blanch or steam corn. Spread in a shallow layer on drying trays.

Indoors. Corn can be dried for 1 to 2 hours in a dryer or 2 to 3 hours in an oven. Start at 135°F.; increase temperature to 160°F. after 1 hour. Reduce to 135°–140°F. for last hour. Stir frequently to prevent corn from clumping and watch carefully during last hour to prevent scorching. Test dryness. Before conditioning the corn, shake several handfuls at a time in a colander or large strainer to remove silk and any other particles. Condition and pasteurize. Store in small amounts in airtight glass containers.

Outdoors. Corn can be dried successfully on trays in the sun in about 6 to 8 hours. Watch carefully during drying and stir frequently to avoid lumps. Condition and pasteurize.

Test for dryness. Brittle, semitransparent.

Garlic

This vegetable keeps well in cold storage and can be frozen easily, so there is usually no reason to dry it. If you want to experiment with it, however, follow the same procedures as for drying onions.

Greens (spinach, kale, chard, mustard)

Selection. Choose unbruised, mature produce.

Preparation. Pick over greens very carefully, removing any damaged leaves. Wash in several changes of cold water. Shake off excess water and place wet greens in a pot. Cover and cook over high heat for 2 to 4 minutes, until leaves have wilted. Spread carefully in single layers on trays.

Indoors. Start at 130°F. When greens look slightly dry, increase temperature to 150°F. As soon as they are nearly dry, reduce temperature to 130°F. and retain until drying is completed. Condition. Pasteurize. Store in cool place. Greens have a tendency to mat, so stir frequently. About 2 to 4 hours in a dryer and 2 to 5 hours in an oven should be adequate. Watch time carefully as it will vary depending upon kind of greens.

Outdoors. Same preparations apply as for indoor drying. Dry in sun for about 8 hours, slightly less time in solar dryer. Condition. Pasteurize. Store in cool place.

Test for dryness. Crumbles easily in hand.

Horseradish

This pungent vegetable can easily be put up in dried form.

Selection. Choose small fresh roots.

Preparation. Peel or scrape to clean. No pretreatment is needed. Grate on coarse blade.

Indoors. Dry at 130°F. in oven or dryer. Gradually increase heat to 150°F. Decrease to 130°F. again as horseradish appears drier. About 1 to 3 hours will be needed for a dryer or an oven.

Outdoors. Horseradish will dry in about 7 to 8 hours if left on trays in the sun. Stir frequently to prevent from sticking to tray.

Test for dryness. Dry, powdery.

Mushrooms

Selection. Persons who gather their own mushrooms should, of course, always take care to choose edible varieties. Fresh mushrooms should have tightly closed, pink gills.

Preparation. Wipe with a damp cloth to clean. Mushrooms benefit from being steamed or blanched for 5 to 7 minutes, depending upon their size. They may discolor if they are not treated with an ascorbic acid bath, but this is not really necessary. Peel large mushrooms; leave small ones whole. Mushrooms should be sliced lengthwise through the stem. The stems are usually much tougher than the caps, so some people prefer to separate them and dry them in two batches. Slice in ⅛-inch pieces. Arrange no more than ¼ inch deep on trays.

Indoors. Heat the dryer or oven to 130°F. and maintain this temperature for about 1 hour. Gradually increase heat to 150°F. and maintain for most of the drying time. When mushrooms begin to appear dried, reduce heat to 130° to 140°F. and watch carefully, removing individual slices as they meet the test for dryness. About 3 to 4 hours is required for a dryer, and 3 to 6 hours for oven-drying. Condition and pasteurize. Store carefully.

Outdoors. Mushrooms can be dried in the sun in about 6 to 9 hours. Shake or stir the trays frequently and watch carefully toward end of drying. Condition and pasteurize.

Test for dryness. Some varieties are very brittle and powdery; others are dry when they feel leathery. A good way to develop an eye and feel for correctly dried mushrooms is to check several of the varieties available in grocery stores.

Okra

This vegetable with its southern origins is an excellent addition to Creole dishes, as well as other chicken and seafood dishes. It is also delicious in soups.

Selection. Fresh okra should be evenly colored and light green; its pointed end should give slightly when pressed.

Preparation. Wash, trim tops, and cut crosswise into ⅛-inch slices. Discard any tough, reedy pods. Blanch in boiling water for 3 to 5 minutes, depending upon size. Arrange in ¼-inch layers on cheesecloth-covered trays.

Indoors. Heat oven or dryer to 130°F. and place okra in it for about 1 hour. As soon as okra takes on a slightly dried appearance, increase the heat to 140°–150°F. Toward the end of the drying time, reduce heat to 130°F. and watch carefully to avoid scorching. Stir frequently during drying. Okra dries in about 2 to 3 hours in a dryer and requires up to 6 hours in an oven.

Outdoors. Follow same procedure as for indoor drying. About 8 to 12 hours of sun-drying is required. Bring okra inside at night or move to a sheltered place. Cool it down before covering with protective sheeting during night. Return to sun the next day. Condition and pasteurize.

Test for dryness. Okra will be tough to brittle.

Onions

Onions are another vegetable that take on their own special character when dried. They enhance soups, stews, pot roasts, and many casseroles.

Selection. Use only strong varieties for best results, such as Sweet Spanish, any of the Creole varieties, Ebenezer, or any large yellow cooking onion.

Preparation. Remove the outer covering and slice uniformly in ⅛-inch-thick slices. Do not boil or steam. Arrange in layers about ¼ inch deep on cheesecloth-covered trays.

Indoors. Start at 140°F. and retain until onions are almost dry. Reduce temperature to 130°F. and watch carefully for scorching during last 45 minutes to one hour of drying. Remove individual rings as they finish drying. One to 3 hours will be needed for a dryer, 3 to 6 hours for an oven.

Outdoors. Onions dried in the sun should be shredded on a coarse grater for faster drying; drying time is about 8 to 11 hours. If drying time extends longer than 1 day, move trays to a protected place in late afternoon. Cool down before covering with protective sheeting. Return to the sun early the next morning. Condition and pasteurize.

Test for dryness. Brittle, crumbles easily in hand when crushed.

Peas

Dried peas are a delicious addition to vegetable soup and also can be used to make a thick soup.

Selection. Choose tender young peas.

Preparation. Shell. Blanch in boiling water for 4 minutes. Spread evenly about ½ inch deep in trays.

Indoors. Start at about 120°F. and maintain for 45 minutes to 1 hour. Increase heat to 150°F. When peas are nearly dry, reduce temperature to 130°F. and maintain. Watch carefully to avoid scorching. Condition. Pasteurize. About 3 hours is needed for dryer or oven.

Outdoors. Arrange in thin layers on trays and dry in sun for 6 to 8 hours.

Test for dryness. Peas should be hard and shriveled.

Peppers, Bell (red and green)

Selection. Choose evenly colored, unbruised vegetables. Best varieties to use are California Wonder, Merimack Wonder, or Oakview Wonder.

Preparation. Wash, pat dry with towel, and trim stem. Cut away interior white sections. Cut into slices of equal size or quarters, whichever seems best for size of pepper. Steam for 8 minutes. Drain and pat dry. Arrange in single layers on trays.

Indoors. Start oven or dryer at 120°F. After about 1 hour, increase heat to 145°–150°F. and retain for about ½ of drying time. Stir frequently during drying. Toward end of drying, reduce heat to 130°F. and watch carefully. Condition. Pasteurize. About 3 to 4 hours will be needed in dryer, up to 6 hours in oven.

Outdoors. Prepare as for indoor drying. Place on trays in the sun. Six to 8 hours will be needed for drying in the sun, slightly less time in a solar dryer.

Test for dryness. Peppers should be brittle.

Peppers, Chili

These dried peppers hold their flavor very well and make an excellent addition to many dishes. They are an important ingredient in Mexican cooking.

Selection. Chili peppers are the long, slender, small ones. They are green when immature and dark red when fully mature. Choose the dark red pods for drying.

Preparation. Wash, pat dry, and remove stem. Cut out soft interior part. Cut into small pieces. No pretreatment is necessary. Arrange pepper bits on trays in ¼-inch layers.

Indoors. Start drying at about 120°F.; increase heat after 1 hour to 145°F. Reduce heat to 130°F. as peppers appear dried. Shake frequently to prevent sticking and to encourage even drying. Test for dryness, condition, and pasteurize. Allow 3 to 5 hours for dryer, 4 to 6 hours for oven drying.

Outdoors. Prepare same as for indoor drying, but cut pods into quarters. String pods about 1 inch apart and hang to dry in the sun. Condition. Pasteurize. Two to 3 days may be required for complete drying.

Test for dryness. Pods will be shriveled and dark red; they will bend easily without snapping.

Pimentos

Follow same procedure as for chili peppers.

Popcorn

Popcorn, which remains plump when dried, is delicious, but not especially easy for the amateur home drier to come by—primarily because it is usually dried in the husk. If you have room to grow your own popcorn, it is quite delicious and worth the small amount of effort required. If you want to do something a little fancier than just letting the corn dry on the stalk in its husk, pick it and peel back the husks. Tie 4 to 5 ears together and hang to dry in the sun. Not only does it look highly creative in an outdoorsy way, but it is a good finishing touch for the drying process.

Test for dryness. Rub off a few kernels and pop them. If you like what you get, remove all kernels and store in tight, moistureproof containers. Popcorn should retain a little moisture—this is what makes it explode when it cooks—so take care not to continue drying it once it has passed the "pop test."

Potatoes

Even though potatoes keep well in root cellar storage, they are interesting dried and can be used in many dishes. They are an excellent thickener in soups.

Selection. Choose white Irish potatoes.

Preparation. Peel and cut julienne style into pieces about ⅛ inch thick by 2 inches long. Blanch for 4 minutes or steam for about 5 to 6 minutes. Pat dry with paper towel. Arrange on cheesecloth-covered trays in layers about ¼ inch deep.

Indoors. Begin at a temperature of about 130°F. and maintain for 1 hour. Raise temperature to 150°F. and maintain until potatoes appear dried. Reduce temperature to 130°F. to finish drying. About 2 to 5 hours will be needed in a dryer and as much as 6 hours in an oven.

Outdoors. Prepare as for indoor drying. About 8 to 12 hours will be needed to complete outdoor sun-drying, slightly less in a solar dryer. Condition. Pasteurize.

Test for dryness. Brittle.

Pumpkins

Strings of drying pumpkins adorned almost all pioneer kitchens. Their uses today are limited primarily to mashing, but they are fun to dry.

Selection. Pumpkins should really be harvested directly from the garden. Choose solid, thick-fleshed ones that have matured to a deep orange.

Preparation. Cut into manageable pieces, remove pith, and shred with coarse blade of food grinder. Steam small quantities for about 5 minutes. Arrange in shallow layers in cheesecloth-lined trays.

Indoors. Start at 120°F. and dry for 2 to 3 hours in dryer or as long as 6 hours in an oven. After the first hour, increase the temperature to about 150°F. and maintain for about ½ the drying time, or until the pumpkin takes on a dried appearance. Drop to 130°F. for last hour of drying. Watch carefully and stir frequently to prevent scorching. Condition. Pasteurize. Indoor room-drying (stringing pieces of pumpkin and hanging them in direct sunlight or near an artificial source of heat) is picturesque, but does not produce especially tasty pumpkin.

Outdoors. Preparation is the same as for indoor drying. About 8 to 12 hours will be needed to complete the drying process. Condition. Pasteurize.

Test for dryness. Pumpkin will have formed into dry chips.

Squash, Winter

Selection. Choose banana or Hubbard squash for drying.

Preparation. Wash, peel, and cut banana squash into strips about ¼ inch thick. Blanch in boiling water for 2 minutes. Drain and pat dry. Cut Hubbard squash into slightly smaller strips. Blanch for 1 minute. Arrange in shallow layers in cheesecloth-covered tray.

Indoors. Start at about 120°F. and maintain for the first hour. Raise temperature to 150°F. and maintain that temperature until squash looks dry. Reduce temperature to 130°F. to finish drying and watch carefully. Remove any pieces that seem to have finished drying and meet test for dryness. Two to 4 hours will be required for dryer and 3 to 5 hours in an oven.

Outdoors. Same preparation is needed as for indoor drying. About 6 to 8 hours is required.

Test for dryness. Squash is tough to brittle when dried properly.

Tomato

Dried tomatoes are quite good stewed or in soups.

Selection. Use home-grown or vine-ripened tomatoes.

Preparation. Dip in boiling water for 30 seconds to loosen skins so they can be easily peeled off. Blanch in boiling water until soft, about 2 minutes at most. Cut tomatoes in half if small; slice them if they are large. Arrange on trays in shallow layers.

Indoors. Start at about 130°F.; raise temperature to 150°F. after about 1 hour. When tomatoes begin to look dry, reduce the temperature to 130°F. and finish there, taking care to avoid scorching. About 3 to 5 hours will be needed to complete process in dryer; 6 to 8 hours is required in oven. Condition. Pasteurize.

Outdoors. Same preparations apply as for indoor drying. About 8 to 11 hours will be needed to complete outdoor drying. Condition. Pasteurize.

Test for dryness. Dried tomatoes are leathery.

Zucchini

Treat the same as Winter Squash.

"6"

COOKING DRIED FRUITS AND VEGETABLES

COOKING DRIED FRUITS AND VEGETABLES

Suggestions for cooking dried fruits and vegetables are based on one simple principle: you must put back into them the water that has been removed through drying. If food is properly dried and cooked, it will return almost to its original size and shape, although the flavor will be distinctly different from that of the canned or fresh varieties. Avoid overcooking dried foods, because their texture and flavor suffer, just as when fresh fruits and vegetables are cooked too long.

Generally about 1 to 2 cups of a dried fruit or vegetable makes about 6 cups or servings when cooked.

There are two ways to prepare dried fruit. First, it may be reconstituted for later use in a baked dish. To reconstitute, place the fruit in a dish and barely cover with boiling water. Cover the dish and set aside for several hours, until the fruit plumps out. Use the liquid in which the fruit has soaked as if it were the natural juice.

The second way is simply to cook dried fruit without reconstituting it. Put it in gently simmering water, cover the pan, and simmer for 10 to 25 minutes, depending upon the kind of fruit and the size of the pieces. A pinch of salt helps to bring out the natural sweetness of the fruit. Add sugar sparingly and only near the end of the cooking, since it will tend to toughen the fibers of the fruit. Less sugar is needed than when cooking fresh fruit, because drying causes some of the starch to convert to sugar. A small amount of fruit juice may help to rejuvenate the taste of the cooked fruit. Remove cooked fruit from heat and cool. It is especially delicious if refrigerated overnight and served with fresh whipped cream.

Finally, remember that dried fruit is excellent eaten as is—with no cooking or special preparation.

Unlike dried fruit, dried vegetables are never eaten uncooked; in fact, they are best when combined with other foods in soups, stews, casseroles, or stuffings.

Greens (including cabbage) need no soaking; simply toss in simmering water and cook for about 5 to 10 minutes, until the taste test shows that they are done.

Other vegetables should be soaked before cooking. Soak them for 20 minutes to 2 hours, depending upon the kind of food and size of pieces, in as much water as they can absorb. Since 1 to 2 cups of dried food converts to about 6 cups of cooked food, it's wise to start soaking in about 2 cups of water. Add more as it is absorbed. Cook the vegetable in the water in which it was soaked, adding still more water as it cooks if necessary. Bring to a simmer slowly, cover, and cook 10 to 30 minutes, depending upon the vegetable. Remember that vegetables have already been cooked somewhat through blanching, and take care not to overcook them.

Vegetables that are to be cooked for several hours in soups need no soaking.

A Word about the Recipes

These recipes show some of the traditional ways—as well as the more unusual ways—that dried foods can be used in daily cooking. Some recipes, such as the Bigos, are not usually made with dried apples, yet the addition of this food in its dried form imparts a new subtle flavor to this delicious stew. This is frequently the case when dried foods are added to dishes that do not normally require them.

It is easy to use dried foods in recipes where they are not called for. If the recipe mentions the food in its fresh form, simply substitute the food in its dried form —using one-third to one-fourth less of the dried food, of course. If you sense that a recipe would work well with a dried food, by all means go ahead and experiment. You will rarely be wrong.

Although smoked meat and fish have a more definitive flavor than dried fruits and vegetables—or even dried meats and fish—many simple casseroles and other cooked dishes are enlivened by the addition of smoked food. This is particularly true if the seasoning is mild and the flavor of the smoked food is allowed to dominate.

In short, although you can and probably will spend many fascinating hours pouring over cookbooks seeking recipes that make use of dried and smoked foods, remember that your own palate is often your best guide to what combinations of food are particularly delicious together.

Fruit leathers—the candy made from the pulp of dried fruit—were used to satisfy the sweet tooth of settlers who found themselves in the American Midwest and West with a plentiful harvest of fruits and little or no sugar, which was both scarce and expensive. Fruit leathers are still sometimes sold in grocery stores, but they are much more fun to make at home.

You need not make these leathers in one step, but can make and freeze the pulp and then make the leathers later. You can experiment with your own flavorings.

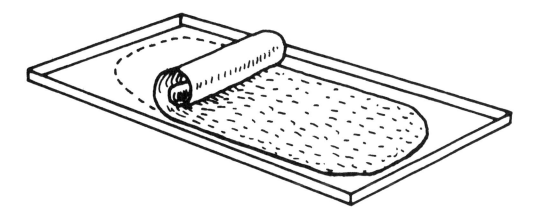

Figure 16. Fruit leather

APRICOT OR PEACH LEATHER

1 gal. pitted apricots or
peaches

1½ cups unsweetened
pineapple juice
3 tsp. almond extract

Place the fruit and the pineapple juice in a large, heavy pot. Cover and cook over low heat. Steam until soft. Drain off the liquids as they accumulate. When the pulp is about the consistency of apple butter, sweeten it to taste with the almond extract. Spread about ¼ inch thick on a lightly oiled cookie sheet. Cover with cheesecloth and place in a warm, dry room or in a 120°F. oven with the door ajar. Leathers take 2 weeks to dry naturally and several hours in an oven. When leather looks dry, peel it in one piece from the pan and place on a rack to finish drying on both sides. Cut into strips and dust with cornstarch, arrowroot, or sugar. Store in a cool, dry place.

The same procedures can be followed for other leathers, but the ingredients vary slightly.

PRUNE LEATHER

1 gal. pitted prunes
1½ cups water

3 tsp. almond extract
honey to taste (added
with almond extract)

APPLE LEATHER

1 gal. apples   *honey*
2–3 cups apple cider

CLAFOUTIS

This pudding is served in French homes as a simple dessert. It is quite easy to make and can be made with dried apples or cherries.

2½ cups dried apples	*½ cup flour*
or cherries	*¼ cup sugar*
boiling water	*2 tsp. vanilla extract*
1½ cups milk	*powdered sugar*
4 eggs	

Put fruit in a pan or bowl and pour boiling water over to cover. Cover and set aside for about an hour, or until fruit has plumped out. Drain fruit and pat dry with a paper towel. Preheat oven to 350°F. and lightly butter a 9-by-9-inch square baking dish. In a blender or mixing bowl, combine milk, eggs, flour, sugar, and vanilla and blend into a smooth batter. Spread the fruit evenly in the baking dish and pour the batter over it. Bake 55–65 minutes, until top is puffed and golden brown. Remove from oven and sprinkle with powdered sugar. Serve warm. Serves 8.

APRICOT-CHERRY SHERBET

¾ cup dried apricots	*1½ Tbs. cognac or light rum*
2 cups water	*4 Tbs. dried cherries*
½ cup superfine sugar	*1 egg white*
1½ Tbs. lemon juice	

Cook apricots in water to cover in a saucepan; drain and remove excess water. Puree in a blender. In a large pot, cook 2 cups water and sugar; when it boils, add pureed apricots and lemon juice. Mix well. Pour into an ice cube tray or mold (or use an ice-cream churn freezer according to manufacturer's instructions). Freeze for at least 6 hours, preferably overnight. One hour before removing from freezer, pour liquor over dried cherries and soak. Immediately before removing from freezer, beat egg white until stiff. Fold egg white and cherries into sherbet. Return to freezer until ready to serve. Serves 4–6.

PEARS STUFFED WITH DRIED FRUIT

The pears for this dish should be ripe, yet firm enough to stand up to the cooking.

5 cups water	4 Tbs. dried apple or other
1 cup sugar	fruit of your choice
½-in. piece of vanilla	3 Tbs. chopped orange peel
4 cloves	2 Tbs. cognac or light rum
6 pears	1 tsp. sugar
4 Tbs. dried white raisins	6 oz. dark sweet chocolate
4 Tbs. dried cherries	3 Tbs. double-strength coffee
	⅓ cup heavy cream

Combine the water, sugar, vanilla, and cloves in a large pot. Bring to a boil, taking care not to spill any of the syrup as it will be hot enough to cause a severe burn. Gently add pears and cook at a simmer for 30 minutes. Remove pan from heat and let cool to room temperature for 20 minutes. Remove pears and cool on a rack. Carefully core pears and cut the bottoms straight across so pears can stand upright by themselves. Combine dried fruits in a small bowl and sprinkle liquor and sugar over them. Let stand for 45–60 minutes.

Carefully stuff pears with dried fruit mixture and place on serving dishes. Just before serving, melt chocolate and coffee in a saucepan. When melted, add heavy cream and stir to mix well. Spoon over pears and serve at once. Serves 6.

CHOCOLATE ROLL STUFFED WITH DRIED FRUIT

½ lb. dark sweet chocolate	4 Tbs. chopped dried cherries,
8 eggs, separated	apricots, or other fruit
1 cup superfine sugar	1–2 Tbs. cocoa
1½ cups heavy cream	

Preheat oven to 350°F. Oil a jelly roll pan and line with waxed paper extending 1 inch beyond length of pan. Put chocolate in a saucepan and melt slowly. Beat yolks and sugar together until pale and light yellow. When melted chocolate has cooled, beat slowly into egg-sugar mixture. Combine thoroughly. Beat egg whites until they form soft peaks and fold gently into mixture. Spread batter evenly in pan and bake 16–18 minutes, until dough is slightly puffed up and surface has a dull finish. Remove from oven, cover with a layer of wet paper towels topped by a layer of dry paper towels, and let stand 10 minutes.

Beat cream, blend in dried fruit, and set aside. Carefully remove paper towels from dough; loosen edges and gently lift the waxed paper to loosen bottom. Use a sifter to sprinkle top of dough with the cocoa. Place a wet, wrung-out dish towel over the surface of the pan, and flip it over quickly. Lift away pan and carefully peel off waxed paper. Spread the whipped cream over the surface evenly. Quickly roll cake toward you. Sprinkle on more cocoa to cover any cracks. Chill until ready to serve. Serves 8.

FRUITCAKE

1½ cups raisins	several tsp. flour
⅔ cup dried chopped currants	2 cups all-purpose flour
¼ cup dried minced apricots	¼ tsp. salt
¼ cup dried minced cherries	1 tsp. baking soda
⅓ cup finely minced candied	1 cup unsalted butter
orange and lemon peelings	½ lb. sugar
½ cup ground pecans	5 eggs

Preheat oven to 275°F. Butter a round 9-inch springform pan and place in it a piece of waxed paper cut to fit the bottom. Butter waxed paper. Combine dried fruits, nuts, and peels in a small bowl and dredge with flour. Sift flour, salt, and soda together. Whip butter until light and fluffy. Add sugar and cream together. Add one egg at a time to butter mixture, beating after each addition. Stir in fruits. Turn into pan and bake about 3 hours, until a toothpick comes out clean. Remove from pan. Serves 6–8.

DATE-RAISIN-NUT BREAD

1 cup boiling water	2 Tbs. butter
1 cup dried dates	1 cup brown sugar, firmly
¼ cup raisins	packed
2 tsp. baking soda	2 eggs, lightly beaten
2 cups sifted all-purpose flour	¼ cup chopped walnuts
1 tsp. salt	

Preheat oven to 350°F. Pour boiling water over dates, raisins, and baking soda in a bowl. Cover and set aside for 1 hour. Sift flour and salt together. Cream butter and brown sugar and stir in eggs. Mix well. Add dry ingredients to creamed mixture, stirring well after each addition. Slowly add date mixture, stirring to mix well after each addition. Stir in nuts. Turn into a buttered loaf pan and bake on the middle shelf of the oven for 1 hour and 10 minutes, or until a toothpick inserted in the bread comes out clean. Cool.

DRIED BAKED FRUIT

2 cups dried peaches, apples,
    prunes
2 cups boiling water
    (optional)
1 cup brown sugar

dash of cloves, cinnamon, or
    other compatible spice
½ cup water
whipped cream

If you wish, plump out the fruit before baking by pouring 2 cups boiling water over it and letting it stand for about 30 minutes. Drain. Arrange the fruit in a baking dish. Sprinkle over brown sugar and spice. Pour water over the mixture. Bake in a preheated 350°F. oven: 25 minutes for peaches, 40 minutes for apples, and 20–30 minutes for prunes and pears. Top with whipped cream. Serves 4–6.

PRUNE WHIP

This recipe is often found in old American cookbooks. It is a delicious way to prepare dried prunes.

2 cups boiling water
1 cup dried prunes, chopped
5 Tbs. sugar
1 tsp. vanilla extract

⅓ cup chopped pecans
6 egg whites
⅛ tsp. cream of tartar
whipped cream

Pour boiling water over prunes; cover and set aside for 20 minutes. Drain and pat dry. Preheat oven to 350°F. Combine prunes with sugar, vanilla, and nuts. In a separate bowl, beat egg whites and cream of tartar until stiff. Gently fold the egg whites into the fruit mixture. Pour into a buttered 2-quart soufflé dish and bake in oven until browned, about 20 minutes. Serve hot, topped with whipped cream. Serves 4.

FRUIT COMPOTE

A fruit compote is always a refreshing dessert. It is also one of the best ways to use dried fruit. The best compotes are made with a combination of fresh and dried fruit.

1 cup sugar
6 cups water
1 cinnamon stick
5 cloves
½ cup dried cherries

¾ cup dried pears
¾ cup dried peaches
1 large orange, peeled and cut
    into fine slices

Combine sugar, water, cinnamon, and cloves in a 4-quart saucepan and bring to a simmer. Carefully (the syrup is very hot and burns the skin easily) add the dried fruit. Cook at a simmer, covered, for 20 minutes. Remove from heat and let fruit cool in syrup for 20 minutes. Place ¼ of the orange slices in a bowl and add about ¼ of the poached dried fruit. Repeat the process until all fruit is used and then pour the syrup over the mixture. Serve at once. Serves 6.

STREUSEL COFFEE CAKE

2 cups flour
3 tsp. baking powder
½ tsp. salt
½ cup sugar
6 Tbs. butter
1 egg
1 cup milk
⅓ cup raisins
⅓ cup minced dried apples

TOPPING:
4 Tbs. butter
½ cup sugar
½ tsp. nutmeg
½ tsp. cinnamon

Preheat oven to 350°F. Sift flour, baking powder, salt, and sugar together into a large mixing bowl. Cut butter into mixture. Add egg and stir until well blended. Gradually add milk, stirring to mix well. Stir in raisins and apples. Pour into a 10-by-14-by-2-inch buttered baking dish. Use a fork to combine the topping ingredients and sprinkle over the batter. Bake 25–30 minutes. Serve warm. Serves 8.

APRICOT BUNS

2 cups boiling water
about 2 cups dried apricots or
  prunes
2 cups flour
1 Tbs. baking powder
½ tsp. salt

3 Tbs. sugar
2 Tbs. shortening
4 Tbs. butter
¾ cup cream
4 Tbs. honey

Pour boiling water over fruit; cover and set aside. Preheat oven to 450°F. Sift flour, baking powder, salt, and sugar together and cut in shortening and 2 tablespoons of butter until pieces of fat are very small. Add cream and work ingredients until they form a soft dough. Knead dough for about 5 minutes, let rest 5 minutes, then roll dough to a thickness of about ¾ inch. Cut into 3-inch-wide circles. Melt remaining 2 tablespoons butter and honey together. Drain apricots and pat dry with paper towel. Make an indentation in each piece of dough and place an apricot in it. Brush the tops lightly with the honey-butter mixture. Place on a buttered cookie sheet and bake in oven 18–20 minutes, until lightly browned. Serve warm. Serves 6.

FRIED PIE

These pastries are traditionally made with dried apples, peaches, apricots, or prunes. They are an excellent dessert and snack food.

1 lb. dried fruit
½–¾ cup sugar, depending
   upon fruit and your taste
vegetable oil

PASTRY:
2 cups flour
½ tsp. salt
1 scant tsp. sugar
⅓ cup shortening
¼ cup water

Put dried fruit in a saucepan with sugar and water to cover; cook over low heat until fruit is tender, about 40 minutes. While fruit cooks, make pastry by blending dry ingredients with shortening in a mixing bowl. Add enough of the water to make a moist dough. Roll out dough to a thickness of ⅛ inch and cut into 3-inch rounds. Cool fruit and mash it. Place a couple of tablespoonfuls of fruit on each pastry round; fold over and seal with water. Heat 2½ inches of vegetable oil to 375°F. in a large pan. Brown one or two of the pies at a time in the oil. Be sure to keep the oil hot or the pies will not fry. Serve hot or cold.

RISOTTO À LA RUSSE

This rice dish, which is excellent with roast chicken, comes from eastern Russia.

¼ cup olive or vegetable oil
2 medium-sized onions,
   chopped
1 large tomato, minced
2 cups long grain rice
4 cups chicken broth

1 cinnamon stick
1 tsp. ground allspice
½ tsp. salt
¼ tsp. freshly ground pepper
½ cup raisins
3 Tbs. butter

Heat oil in a skillet and sauté the onion in it. Add tomato and cook for 1 minute. Add rice and cook, stirring constantly, until rice is translucent and well coated with the oil. Pour in broth. Add seasonings and raisins; cover and cook 20–30 minutes, until all broth has been absorbed. Stir in the butter. Remove cinnamon stick. Serves 8.

SAUSAGE-APPLE STIR-FRY

1 pkg. link breakfast sausage
½ lb. dried apples, preferably
   rings

4 Tbs. water
¼ tsp. ground ginger
salt and pepper to taste

Place sausage in a cold skillet; turn heat on high. When sausages begin to cook, reduce heat to medium-high and stir sausage occasionally to brown evenly. When browned, add apple rings and water. Cover and cook 5 minutes over low heat. Stir in ginger, salt and pepper to taste. Cook uncovered 1 minute. Serve at once as a light supper or breakfast dish. Serves 4-6.

CHICKEN STUFFED WITH DRIED FRUIT

¼ cup butter	2 Tbs. raisins or dried
¼ cup onion, minced	currants
½ cup long-grain rice	1 tsp. salt
1 cup chicken stock	1 tsp. cinnamon
½ cup dried apricots, finely	¼ tsp. freshly ground pepper
chopped	1 3½-lb. frying chicken
	¼ cup honey

Melt 2 tablespoons butter in medium-sized saucepan. Add onion and cook until soft. Add rice, stirring so every grain is coated with the butter. Cook uncovered for 5 minutes. Pour in chicken stock; add apricots, raisins, salt, cinnamon, and pepper. Cover and cook 15–20 minutes, until liquid is absorbed. Remove from heat. Preheat oven to 375°F. Wash chicken and pat dry. Melt remaining 2 tablespoons butter and honey together. Stuff chicken with rice mixture and skewer or sew up so rice will not spill out. Place chicken in roasting pan in oven and roast 15 minutes. Reduce the heat to 325°F. and coat chicken with the honey-butter mixture every 15 minutes during cooking. Roast 1½ hours or until juices run yellow and leg moves easily. Serves 4.

TOKANY

Tokany is one of the great Hungarian stews, and there are many versions of it. This one makes use of dried mushrooms, which have an exotic flavor all their own, and dried marjoram.

1 cup dried mushrooms	1 tsp. salt
4 slices bacon	1 tsp. freshly ground black
1 lb. lean pork, cut into	pepper
3-by-1-in. strips	¼ tsp. dried marjoram
1 cup water or dry white wine	1 cup sour cream
1 Tbs. sweet Hungarian	
paprika	

Soak mushrooms for 1 hour in water. Squeeze dry and set aside. Heat bacon; add meat and sauté until each piece is browned.Add water or wine and seasonings. Cover and cook until meat is tender, about 1½–2 hours.

Add mushrooms during last 30 minutes of cooking. Immediately before serving, add sour cream in large spoonfuls, stirring to blend thoroughly after each addition. Heat through, but do not boil. Serve over noodles and with a green salad. Serves 6.

DRIED ZUCCHINI AND CHEESE CASSEROLE

2 cups boiling water
2 cups dried sliced zucchini
8–10 slices Swiss cheese
salt and pepper to taste

1 8-oz. can tomato sauce
1 cup bread crumbs
¼ cup Parmesan cheese

Pour boiling water over zucchini; cover and let stand for 1 hour. Drain zucchini and pat dry with paper towel. Preheat oven to 350°F. Arrange a single layer of zucchini in a buttered baking dish. Cover with a layer of Swiss cheese. Sprinkle with salt and pepper. Spoon over a thin layer of tomato sauce and top with bread crumbs. Repeat until all zucchini and cheese slices have been used, finishing with a light layer of bread crumbs. Sprinkle Parmesan cheese over casserole. Bake for 25 minutes, or until lightly browned. Serves 4-6.

MINESTRONE

This famous Italian vegetable soup is delicious when made with dried herbs and vegetables.

2 slices bacon
2 Tbs. butter
4 Tbs. dried onions
⅓ cup chopped fresh leeks
1 carrot, peeled and chopped
2 garlic cloves, minced
4 cans chicken stock, mixed
  with equal amount of water
1 cup dried cabbage
2 cups dried zucchini

1 1-lb. can Italian tomatoes
  or 5 medium-sized fresh
  tomatoes, chopped
1 large Idaho potato, peeled
  and sliced
1 1-lb. can white beans
2 Tbs. dried basil
1 bay leaf
salt and pepper to taste
1½ cups small pasta or rice
1 cup freshly grated Parmesan
  cheese

Heat bacon and butter in a Dutch oven or soup kettle until very hot. Add onions and leeks and sauté for 2 minutes, stirring constantly. Add carrot and garlic and sauté for 5 minutes, stirring occasionally. Add diluted chicken stock and all remaining ingredients except pasta or rice and

Parmesan cheese. Partially cover and simmer 40 minutes. Add pasta or rice and simmer uncovered 10 minutes. Serve soup hot and pass Parmesan cheese as a topping. Serves 8-10.

CREAMED VEGETABLE SOUP

Dried vegetables are excellent in a cream soup, particularly when cooked in the broth they have soaked in.

4 Tbs. butter
1 large onion, diced
1 carrot, diced
1 potato, diced
1 tsp. salt
½ tsp. freshly ground pepper

¾ cup dried vegetable of your choice, soaked in 2 cups water for several hours
3 cups chicken broth or water
1 cup light cream

Melt butter in a soup pot, and sauté onion and carrot for 5 minutes. Add all remaining ingredients except cream. Simmer, partially covered, for 40 minutes. Remove from burner and cool. In small quantities, put soup through blender and pour into tureen. Stir in cream. Serve chilled or hot. Serves 6.

SHAKER DRIED CORN

2 cups boiling water
1 cup dried corn
2 tsp. sugar

½ tsp. salt
3 Tbs. butter
⅔ cup light cream

Pour boiling water over the corn and let it stand 1 hour. Stir in the sugar, salt, and butter. Cook, uncovered, over low heat about 30 minutes, until most of the liquid is absorbed. Stir in cream and heat through. Serves 4.

VEGETABLE FLAN

4 Tbs. butter
1 cup water
1 lb. mixed and diced dried vegetables, such as eggplant, carrots, zucchini, onion, green beans and green pepper
1 lb. red potatoes, peeled and diced

1 dried bay leaf
1 tsp. dried basil
1 tsp. fresh parsley, chopped
1 cup heavy cream
4 eggs, separated
½ cup grated cheese
2 Tbs. bread crumbs

In a large saucepan, bring butter and water to a boil. Add vegetables, potatoes, and herbs. Cover and cook until soft, about 40 minutes. Stir in cream and heat through. Remove from burner and cool. Add egg yolks one by one. Beat egg whites until stiff; fold into mixture. Fold in cheese. Pour into an oiled pan, sprinkled with the bread crumbs. Cover with buttered waxed paper and bake 45 minutes at 350°F. Serve hot. Serves 6.

RATATOUILLE

This version of the French vegetable stew takes less cooking time than versions using fresh vegetables.

6 cups boiling water
2 cups dried eggplant
2 cups dried zucchini
1 cup dried onion
⅓ cup dried green pepper
3 cloves garlic, minced
¼ cup vegetable oil

1 Tbs. olive oil
1 1-lb. can tomatoes
1 bay leaf
4 Tbs. fresh chopped parsley
1 tsp. dried basil
salt and pepper to taste

Pour boiling water over dried vegetables; cover and set aside for about 1 hour. Melt vegetable and olive oils in a large pot until hot. Add drained dried vegetables, tomatoes with their liquid, and seasonings. Bring to a simmer; reduce heat, cover, and cook for 45 minutes, until vegetables are tender and juices have thickened slightly. Serves 6.

CHILI CON CARNE

1½ cups boiling water
½ cup dried onion
¼ cup dried green pepper
1 lb. ground beef
1 1-lb. can tomatoes
1 6-oz. can tomato paste

1 1-lb. can red kidney beans
1 Tbs. chili powder (or more to taste)
½ tsp. salt
¼ tsp. freshly ground pepper

Pour boiling water over onion and green pepper; cover and let stand for about an hour. Brown meat in a large skillet or pan. Add dried vegetables, water, tomatoes, tomato paste, kidney beans, chili powder, salt, and pepper. Bring to a boil and cook, uncovered, at a simmer for about 1 hour. Add more water if necessary. Adjust seasonings. Serves 6.

BIGOS (HUNTER'S STEW)

Bigos is a traditional national stew of Poland, which, like all traditional dishes, has as many versions as there are cooks.

1 cup dried mushrooms	1 large can tomatoes
1 cup warm water	1 cup water
4 Tbs. vegetable oil	2 tsp. Bavarian-style mustard
2 large onions, sliced	1 tsp. salt
2 cups sliced cabbage	½ tsp. freshly ground pepper
1 lb. chopped lean beef	1 lb. Polish smoked sausage,
1 lb. chopped pork	cut into bite-sized chunks
3 lbs. sauerkraut, rinsed and	6 red potatoes, peeled and cut
drained	into halves or quarters
2 cups dried apples	

Soak mushrooms in warm water for 30 minutes; drain, reserving liquid. Squeeze mushrooms to remove excess liquid. Heat oil in a large pot and sauté onion and cabbage until tender, about 5 minutes. Push to one side and add beef and pork. Sauté until light brown. Add all the remaining ingredients except sausage and potatoes. Also add reserved broth from mushrooms. Cover and cook over low heat for about 2½ hours, until meats are fork-tender. Add potatoes during last 45 minutes of cooking; add sausage during last 30 minutes of cooking. Serve hot. Serves 8-10.

CHILLED FRUIT SOUP

2 cups boiling water	1 stick cinnamon
1 cup dried sliced peaches	2 cups water
¾ cup sweet dried cherries	½ cup sugar
6 thin slices orange	¼ tsp. salt
4 Tbs. lemon juice	1½ tsp. cornstarch
1 Tbs. grated lemon rind	whipped cream

Pour boiling water over the dried fruit; set aside for 40 minutes to plump out fruit. Place orange slices, lemon juice, lemon rind, cinnamon stick, water, sugar, and drained dried fruit in a large saucepan. Bring to a boil and simmer for 4 minutes, uncovered. Carefully stir in salt and cornstarch, which has been mixed with 1–2 tablespoons water. Cook, stirring, until mixture becomes clear. Adjust seasonings to taste. Remove cinnamon stick. Chill thoroughly before serving. Pass whipped cream topping. Serves 4-6.

HEARTY DRIED VEGETABLE SOUP

2 cups boiling water
½ cup dried green peppers
½ cup dried onion
½ cup dried carrots
¾ lb. ground beef or leftover
  chopped beef
1 1-lb. can tomatoes

2 cans beef stock mixed with 2
  cans water
½ tsp. salt
1 tsp. dried tarragon
1 bay leaf
1 cup cooked long-grain rice

Pour boiling water over dried vegetables and set aside for 1 hour. Brown ground beef in a large skillet. Transfer to a soup pot; add vegetables and their broth, tomatoes, diluted beef stock, salt, tarragon, and bay leaf. Cover and cook for 40 minutes at a slow simmer. Uncover and stir in rice; heat through. Serve hot. Serves 6-8.

MARINATED GREEN BEANS

Green beans, which are among the tougher dried vegetables, adapt well to this treatment if they are of high quality.

4 cups water
2 cups dried green beans
½ tsp. salt
¼ tsp. freshly ground pepper

½ tsp. dried mustard
1 tsp. Dijon mustard
2 Tbs. tarragon vinegar
5 Tbs. olive oil

Place water and beans in a saucepan and boil for 40 minutes or longer, until beans are fairly tender. Drain and pat dry with paper towels. In a small bowl, place salt, pepper, mustards, and vinegar. Add oil. Use a fork or small wire whisk to mix well. Pour over green beans and marinate in refrigerator for 4 hours before serving. Remove from refrigerator about 20 minutes before serving and toss gently. Serves 4.

APICIUS' HAM WITH DRIED FIGS

Apicius was a Roman gourmet to whom the first-century cookbook *De Re Culinaria* is generally attributed. This recipe was adapted from this ancient cookbook.

1 cup hot water
12 dried figs
7-8 lbs. smoked ham

¼ cup brown sugar
pinch of ground cloves
¼ cup fig juice

Soak figs in hot water for 30 minutes. Drain and cut almost in quarters; flatten each fig into a four-leaf clover shape. Set aside. Rub ham with sugar, to which cloves have been added. Bake 18–20 minutes per pound in a 300°F. oven. When sugar has melted, after about 20 minutes, add fig juice. Baste ham every 20 minutes. Add more fig juice if necessary. Thirty minutes before ham is cooked, remove from oven and decorate with figs, attaching them with toothpicks. Complete cooking. Serve hot or warm. Serves 8.

EGGPLANT CAVIAR

This appetizer, which is Eastern European in origin, is an excellent way to use dried eggplant and onion.

½ cup dried onion
¾ cup dried minced eggplant
5 Tbs. olive oil
½ cup finely chopped green pepper
1 tsp. finely chopped garlic

3 medium-sized tomatoes, seeded, peeled, and chopped fine
½ tsp. sugar
1-1½ tsp. salt
½ tsp. freshly ground pepper
3 Tbs. lemon juice

Place onion and eggplant in a saucepan with water barely to cover and cook over low heat for 20 minutes. Drain and squeeze dry. Place mixture in a saucepan or small skillet and add oil, reserving 2 tablespoons. Cook until heated through. Add green pepper and cook 5 minutes more; add garlic and cook 5 minutes more. Turn mixture into a mixing bowl, add tomatoes and seasonings, and mash with a wooden spoon until almost pureed. Add remaining 2 tablespoons oil to saucepan or skillet and return mixture to it. Cover and cook for 1 hour over very low heat, until all moisture has evaporated. Adjust seasoning. Chill before serving on crackers or dark bread. Serves 6.

CLAM-CHEESE CANAPÉS

1 cup grated cheddar cheese
1 8-oz. can minced clams
1 Tbs. dried thyme

2 Tbs. dried onion, soaked in water for 1 hr. and squeezed dry
8 slices bread, trimmed of crusts and quartered

Mix together all ingredients and spread on small toast rounds. Broil 5 minutes, until cheese has melted. Serves 4.

7

DRYING MEAT AND FISH

DRYING MEAT AND FISH

Generally, outdoor air drying of meat and fish is not practical for the amateur food drier these days. First, successful air-drying of these foods requires an expertise that most home driers have not had an opportunity to acquire. Second, there simply are not enough areas untouched by civilization to make drying outdoors safe for these foods. There are, however, ways that meat and fish can still be dried at home. Ironically, the methods used are among the oldest known to us.

Dried meat and fish have a long and honorable history, having been the first "convenience" foods as well as the first foods that were portable and safe enough for persons on long journeys to carry with them.

Pemmican and Charqui

Two of the better known traveling meat products originated in the Americas. The first, called pemmican by the Cree Indians who gave it its name, can still be made easily today. Basically, pemmican is meat dried over an open fire or in the sun. Originally, antelope, buffalo, and other game were the sources of pemmican; today, lean beef works quite well. When the meat was thoroughly dried, it was pounded between two stones until it broke down to a powdery consistency. The Cree combined the powdered meat with fat, bone marrow, and dried wild cherries. Pemmican was stored in a rawhide bag, which was then sealed with tallow. It was the traditional sustenance for early explorers, such as Alexander Mackenzie, who in 1793 was the first white settler to travel from coast to coast in North America.

A modern-day version of pemmican can be made as follows:

Cut 2 pounds of very lean beef into thin strips. String the strips on steel wire and hang them in the sun where insects and dirt cannot reach them. Depending upon the amount of humidity, the strips should take about twenty-four hours to dry. If you live in a humid area or cannot protect the drying meat as suggested, smoke it over hickory or ash coals that are glowing but have not flamed for a few hours. Insect pests will not bother this meat. The meat can also be dried in an oven. Preheat it to 150°, place meat in the oven, reduce heat to 120°. Let meat remain in oven for five hours, then turn over and allow to dry for another four hours. Meat should be dark and shriveled when done. Melt an equal amount of fat, and while it is melting, pound the meat into a coarse powder. Mash a cup or so of mixed dried fruit (currants, apricots, raisins, or prunes work well together or alone) and add to the beef. Slowly add enough hot fat and work the mixture with two wooden spoons until it takes on a doughlike consistency. Form the pemmican into sausage-shaped loaves about 6 to 7 inches long. Wrap in a single layer of muslin and dip in melted paraffin to seal it and make it watertight. Store in a dry place.

The second traveler's food, called charqui, was developed by the Indians of Peru. It, too, was originally made from game, but once cattle became plentiful in the Americas, people quickly switched to beef. Today, charqui is sold in grocery stores and other food shops as a party snack—under the name "jerky," which is a derivative form of the Indian word.

An interesting version of oven-dried jerked meat can be made in the following way:

2.lbs. lean beef   *½ tsp. onion powder*
5 Tbs. soy sauce   *1 generous tsp. salt*
½ tsp. fresh ground pepper

Be sure that all traces of fat are trimmed off beef. Preheat oven to 150°F. (no higher than 200°F.). Combine all ingredients except meat in a large bowl. Cut meat into very thin slices, cutting with the grain if possible. Marinate meat about 1 hour, until the liquid has been absorbed. Remove meat from marinade and pat dry with paper towels. Place meat slices on oven racks so they are not touching each other. Dry in oven for about 5–7 hours, until meat takes on a brown color and a hard, dry appearance and texture. Let cool to room temperature and store in an airtight jar. *Note:* Once you have made jerky a few times, you may want to experiment with different spices and make your own combinations.

Hunters still cherish jerked venison, and there are probably as many versions of this as there are hunters. Should you want to dry venison or beef over an open fire, here is the method:

Cut the meat with the grain into strips about 2–2½ inches thick. Dip the venison into boiling brine, made with the standard ingredients: 4 pounds salt, 1½ pounds sugar, 1½ ounces saltpeter, and 2½ gallons water. String the meat on a drying rack made from wood you have gathered. Build a fire with hardwood, and let it burn low. Place the smoking rack over the fire and sear the meat. Do not let the fire blaze, or the meat will cook. When the meat has seared, dry it in the sun until it is hard to the touch. Eat at once or store in an airtight jar or container.

Drying Fish

It is almost impossible to air-dry fish in North America because the weather is neither suitable nor consistent enough. Also, air-drying fish that has not been salted first requires a great deal of skill if it is to be done with any measure of safety. There is no special skill involved in drying fish that has been salted, however, and you may want to try it.

One of the best kinds of dried fish is called rackling. For many centuries, Scandinavian fishers have prepared it on the beaches where they bring their daily catches. Fish with a fat content of 5 percent or less are easiest to dry, so lean fish such as flounder, haddock, halibut, and rock cod are especially well suited to this particular drying technique.

No matter what type of fish you are drying, buy the freshest fish you can find, or dry your own catch. If you buy at the store, ask your fishmonger to split the fish lengthwise and remove the backbone, leaving the collarbone intact. (The collarbone keeps the fish intact as it dries.) Wash thoroughly until all traces of blood have been removed.

Soak the fish in a brine solution made with 5 pounds of salt and 2 gallons of water. Cure in this way for about an hour. Then remove the fish from the brine cure and pat dry.

Hang the rackling in a shady place by the method shown on the next page in Figure 17 and let dry for about 2 weeks. Fish dried outdoors are always dried in the shade; they spoil too quickly to risk sun-drying. You can use an electric fan to create a breeze, if necessary. Rackling can, and most frequently is, eaten as is, but it can also be soaked for a few hours and then cooked into a variety of fish dishes, such as fishcakes, fish loaf, or creamed dishes.

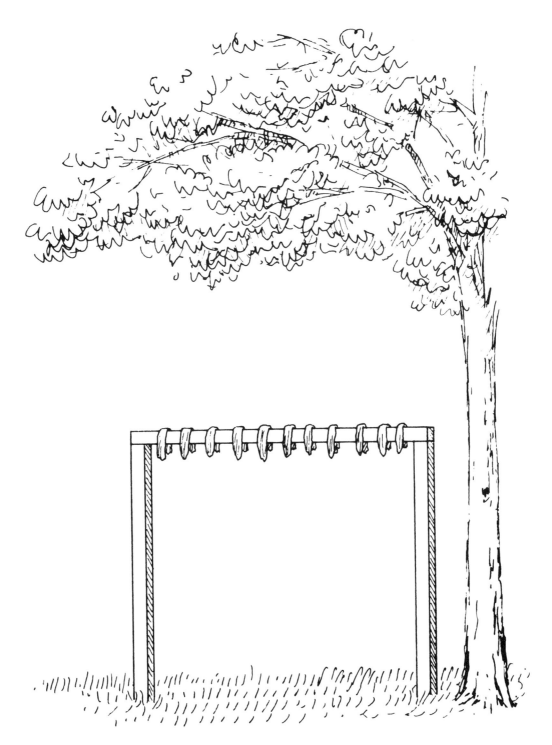

Figure 17. Fish drying in shade

Drying Shrimp

Shrimp can be easily dried at home. It makes an excellent addition to Creole dishes and is frequently used in Oriental cookery. It is also delicious when mixed with sour cream or cream cheese to make an appetizer dip.

To dry shrimp, clean thoroughly and allow to drain dry. Boil the shrimp for 10 minutes in a brine solution made from ¾ cup salt and 1 quart of water. Pat dry with paper towels and remove any shells that have not already separated from the shrimp. Arrange the shrimp no more than 1 inch deep on drying trays and place in a sunny place to dry. Shake or stir the shrimp every 30 minutes to insure even drying. In a climate with low humidity, about 3 days will be needed to dry shrimp. In an area with high humidity, up to a week may be needed. When the sun is no longer hot enough to dry the shrimp in the late afternoon, remove them to a dry, well-ventilated place until morning. Do not cover the shrimp during the night or they may sour. Return the shrimp to the sun every day until drying is completed. Shrimp shrink considerably and feel slightly hard or even powdery when dried.

8

HOW TO
BUILD A SMOKEHOUSE

HOW TO
BUILD A SMOKEHOUSE

The principle of smoking meat, fish, and poultry is old—indeed, it probably was discovered quite by accident when prehistoric people noticed that meat prepared over a smoking fire had a unique taste, and more important, kept for several weeks without spoiling.

Smoked food has always played an important role in daily life, particularly in times when other methods of preserving food were not yet available or recommended.

Imperial Rome had problems preserving food because of the city's astounding size—it was a quarter the size of modern-day Paris. Fresh food was stored in central warehouses, where it was not always convenient to retrieve it for every meal. In addition, people had begun to notice that "fresh" food did not always settle well on their stomachs—particularly when the weather was hot. As fresh food became more and more suspect, people began turning more frequently to salted and smoked fish and meat. They became an important part of the daily diet.

Surely one of the major tasks of the medieval cook was to disguise the taste of slightly rancid meat or fish. Again, people simply learned to turn to smoked foods—if they could afford them. In medieval times, meat could be gotten fairly cheaply—it was the salt, and possibly some spices such as peppercorns and cloves to add an exotic flavor, that cost dearly.

During the Crusades, however, salted and smoked meat and fish were the lifeblood of travelers, and so they have more or less remained today. Smoked food has been favored for many years by mountain climbers and Arctic explorers—and campers. It is the perfect portable food.

The spread of Christianity and its 40-day period of Lenten fasting probably did more for salted and smoked fish than anything else. In England, hanging was the penalty for eating meat during Lent until the sixteenth century, so people who lived inland where fresh fish was scarce quickly learned to use dried and smoked fish.

In the United States, the smokehouse was an important part of farm life from colonial days. "A meathouse is one of the first houses built; hung on all sides with chines, middlings, joles, and hams; perhaps finer flavored for having run wild." So recorded New England poet Henry Knight, in his book *Letters from the South and West*, written in 1818. At this time, tradition held that one found good beef and bad bacon north of the Potomac, and good bacon and bad beef south of the Potomac. Today's driers should take some heed of this information: good bacon or any other smoked food isn't a matter of luck. It requires good techniques and high-quality foods to start with.

Regardless of what Southern-style beef tasted like, the famous Smithfield hams, which grace tables even today, had their reputation for excellence as early as 1639, almost 100 years before the founding of the town of Smithfield, Virginia.

As white settlers moved westward, smokehouses continued to be one of the most important buildings on any farm. Today, the fields of the Midwest are dotted with small, gray fieldstone buildings that once held a winter's worth of fine eating.

The process of smoking meat and fish is not especially difficult. The food is first cured in salt and then smoked until it takes on the traditional smoked flavor, texture, and color. Smoking also lowers the moisture content, thus acting as a preservative. It slows chemical reactions such as bacterial growth and oxidation.

Before you can smoke anything, however, you must have a smokehouse—or some other container suitable for smoking foods. This can be as simple as a barrel smokehouse, an old icebox rigged for use as a smokehouse, or a more elaborate wood or stone smokehouse.

Building a Barrel Smokehouse

A barrel or drum smokehouse will successfully smoke 2 or 3 roasts, hams, shoulders, or slabs of bacon at a single time.

If you can find an old wooden barrel, the type that was often used for curing pork, it will make an excellent container for this purpose. Another excellent container is an oil drum, which needs only a little preparation to make it ready for smoking.

Remove one end of the oil drum if this has not already been done. Buy a drum that has been cleaned so that there is no oil residue. Scrub the drum with a strong detergent and lots of hot water. As an extra precaution, rinse it with boiling water. Dry in the air.

A barrel or drum smokehouse is intended for outdoor use. Besides the drum, you will need:

- 10 to 12 feet of stovepipe
- a stovepipe elbow piece
- 2 or 3 1-inch dowels, slightly wider than the diameter of the drum or barrel
- a roll of steel wire
- cheesecloth or other loosely woven material, enough to cover the top of the container
- a piece of 1-inch-thick wood large enough to cover the barrel or drum
- 2 1-by-1-inch wood strips slightly longer than the barrel opening
- a piece of sheet metal, approximately 2-by-2-feet square, large enough to cover the fire pit
- a clip-on oven, candy, or meat thermometer

You will also need equipment for cutting a hole in the barrel or drum, a shovel, and a strong digging arm.

Figure 18 shows how this type of smokehouse is put together. Begin with the fire pit. You may line it with bricks or any other similar substance, but this is not necessary. The pit should be 1½ to 2 feet deep and about 2 feet wide. Dig the trench at a 30° angle so there will be enough draft on the fire to move the smoke into the smokehouse. The trench should be 10 to 12 feet long, so the smoke will cool before it reaches the food.

When the trench has been dug, assemble the stovepipe in it, using the elbow piece at the end where the barrel will be placed. Cut a hole the size of the stovepipe's diameter in the barrel or drum and fit this container over the elbow piece of stovepipe. Build a small mound of dirt around the drum when it is in place to make it even more airtight.

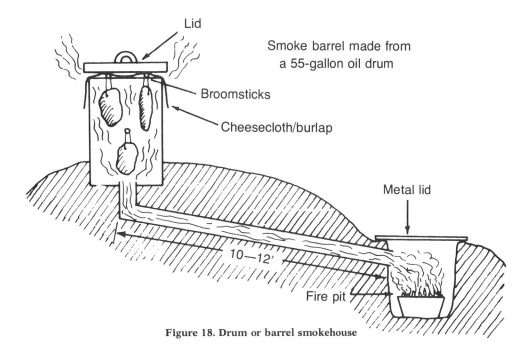

Figure 18. Drum or barrel smokehouse

Using the Barrel Smokehouse

The first step is to bring the smokehouse to the proper temperature. Build a fire and make sure it is going strong. Hook the thermometer in place in the barrel or drum. Regulate the temperature by opening and closing the metal piece over the fire pit. A stone or brick can even be used to lift the piece farther off the ground, if needed.

Attach the food by steel wire to the dowels and put them in place over the open end of the drum. A third piece of food may be strung in the center by steel wire. In addition to holding the food, the dowels help to provide ventilation. If more air is needed, position the wood strips at right angles to the dowels on top of them.

Building a Smokehouse from a Refrigerator

If you have or can buy a used or even damaged refrigerator, it can be converted to a simple smokehouse. One advantage is that this type of smoking container can be used indoors. The disadvantage is that it is difficult to regulate the temperature, for reasons that become obvious when you study Figure 19: the source of heat is not very far from the container, and since the door must be kept closed, it is difficult to regulate the temperature that way.

Only a few simple items are needed to convert an old refrigerator to a smokehouse:

•a wood or cardboard box, approximately 2 feet square
•2 1-inch dowel rods
•steel wire
•a metal dish suitable for burning fuel
•a sheet of metal the size of the firebox
•2 butterfly vents
•putty

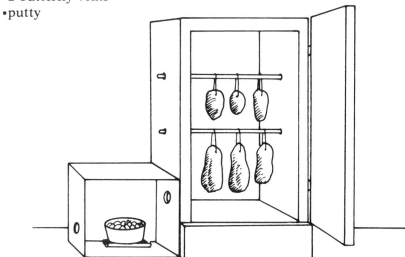

Figure 19. Refrigerator smokehouse

To prepare the refrigerator for its new purpose, strip it bare inside, taking special care to remove any metal pieces that might be galvanized. Plug any openings so the refrigerator is as airtight as possible. Now proceed to make a few new openings: cut a 5- to 7-inch hole in the lower side where the firebox will go. Install 2 butterfly vents, 1 on the firebox and 1 high on the opposite side. These vents will be used to create a draft during smoking. Drill holes in the sides of the refrigerator and install dowel rods. Seal the spaces for the dowels on the outside with putty.

To make the firebox, place a sheet of metal on the floor or whatever surface the smoke container is resting on. (If it is outdoors, this fire precaution may not be necessary.) Use a wood or heavy cardboard box as a cover. Burn the fuel in a fireproof metal dish.

The smoking procedure is the same as for the barrel or drum smokehouse.

Building a Wood-Frame Smokehouse

At first glance, a wood or stone or brick smokehouse may appear to be a difficult undertaking. It isn't. In fact, it is simply an enlarged version of the barrel or drum smokehouse, as careful study of the plans on the following pages shows. Materials can be rather expensive for this type of smokehouse, so it probably should be reserved for serious smoked-food aficionados. On the other hand, this smokehouse can be built from recycled materials as a means of cutting cost—or even from fieldstone, which usually comes cheap if you don't count the labor involved in transporting it. In addition, the fruits of one's labor in this kind of smokehouse can keep an average-sized—or even a large—family well supplied with an assortment of smoked meat, fish, and poultry.

Figure 20. Wood-frame smokehouse

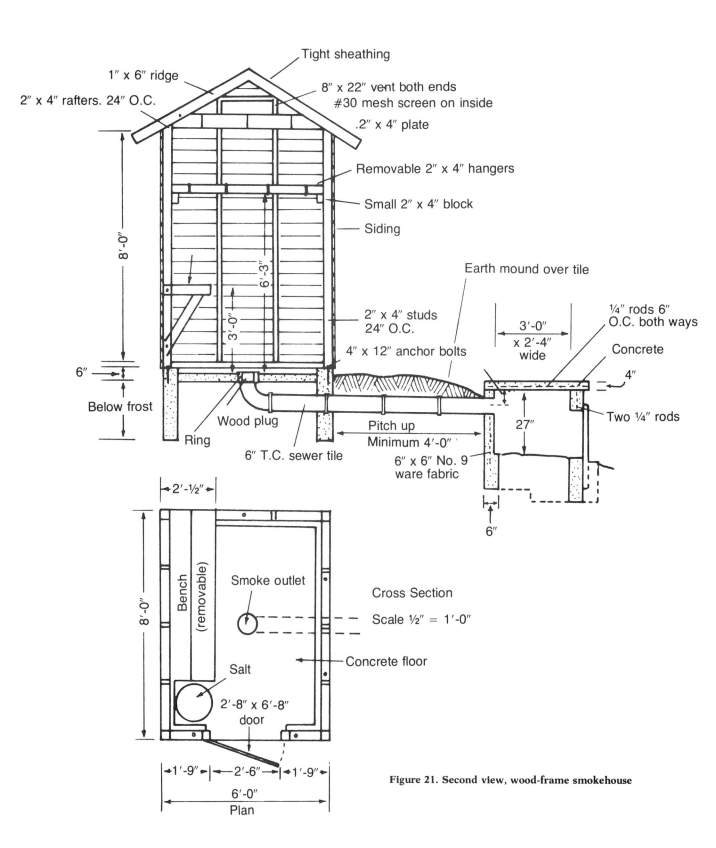

Figure 21. Second view, wood-frame smokehouse

Tight sheathing

1″ x 6″ ridge

2″ x 4″ rafters. 24″ O.C.

8″ x 22″ vent both ends
#30 mesh screen on inside

.2″ x 4″ plate

Removable 2″ x 4″ hangers

Small 2″ x 4″ block

Siding

8′-0″

6′-3″

3′-0″

Earth mound over tile

2″ x 4″ studs
24″ O.C.

4″ x 12″ anchor bolts

3′-0″
x 2′-4″
wide

¼″ rods 6″
O.C. both ways

Concrete

4″

6″

Below frost

Wood plug

Ring

6″ T.C. sewer tile

Pitch up
Minimum 4′-0″

6″ x 6″ No. 9
ware fabric

27″

Two ¼″ rods

6″

2′-1½″

8′-0″

Bench
(removable)

Smoke outlet

Salt

2′-8″ x 6′-8″
door

1′-9″   2′-6″   1′-9″

6′-0″

Plan

Cross Section

Scale ½″ = 1′-0″

Concrete floor

The plans in Figure 21 are self-explanatory, but a few points deserve special mention:

The trench through which the smoke is funneled should be at a 30° angle, as was suggested with the barrel or drum smokehouse.

This smokehouse must be set on a waterproof and frostproof foundation. Concrete works best, but other suitable materials may be used.

The fire pit suggested in these drawings, which were furnished by the U.S. Department of Agriculture, is rather elaborate, as befits a permanent smokehouse. If you do not want to go to this much trouble, you can pattern the pit after the barrel smokehouse in Figure 18.

Two-by-four strips of wood are used to hold the food. It should be hung from the wood strips with steel wire.

Smoking in a smokehouse is very similar to smoking in the barrel or drum version. The fire should be going strong before the food is put in. Put in as much food as the smokehouse can hold, provided it does not touch the wall or any other piece of food. You will soon find yourself adding extra hooks and strips of wood as you become more sure of the amount the smokehouse can handle.

Install a thermometer on the door of the smokehouse or, ideally, install a thermometer that lets you read the inside temperature without opening the door.

The following list of supplies is needed to build the wood-frame smokehouse:

CONCRETE 1:3:5 MIX
•10 bags cement
•1 cu. yd. sand
•1½ cu. yds. gravel

LUMBER
•3 pieces 2″ × 6″ × 6′ crossties
•3 pieces 2″ × 4″ × 12′ header, bench, and hangers
•12 pieces 2″ × 4″ × 10′ sills, studs, and rafters
•23 pieces 2″ × 4″ × 8′ sills, studs, plates, and hangers
•2 pieces 1″ × 10″ × 8′ bench and vent doors
•2 pieces 1″ × 6″ × 10′ ridge and trim
•2 pieces 1″ × 6″ × 7′ door battens
•6 pieces 1″ × 4″ × 10′ corner boards and trim
•6 pieces 1″ × 4″ × 8′ trim
•6 pieces 1″ × 6″ × 7′ T. & G. boards for door
•275 ft. B.M. 6-inch drop siding
•120 ft. B.M. roof sheathing
•Roofing or asphalt shingles to cover 100 sq. ft. roof area

MISCELLANEOUS
•7½-by-12-inch anchor bolts, with nuts and washers
•1 pair 8-inch T hinges
•1 safety hasp
•2 pair 2-by-2-inch hinges, for vent doors
•1 piece metal flashing—6 by 40 inches
•4 lin. ft. #30-mesh wire screen 10 inches wide

•4 lengths 6-inch T.C. sewer pipe
•1 piece 6-inch 90° T.C. elbow
•nails, hanging hooks, and paint

To build the firebox, you will need:

CONCRETE 1:2:3 MIX
•6 bags cement
•¼ cu. yd. sand
•1/3 cu. yd. broken hard brick, gravel, or stone (brick is more heat resistant)

MISCELLANEOUS
•10 lin. ft. 6-by-6-inch No. 9 wire fabric, 30 inches wide
•6 pieces ¼-inch steel rods, 42 inches long
•10 pieces ¼-inch steel rods, 32 inches long
•2 pieces 1¼-inch pipe, 36 inches long
•1 piece 24-by-32-inch metal, sliding door

Choosing the Right Fuel

Hardwoods such as hickory, maple, oak, ash, birch, and dry willow are the best for smoking either meat, fish, or poultry. Dried corncobs are also an excellent source of fuel. Do not use a soft or resinous wood, as it will impart a distasteful flavor to the food. Also to be avoided are artificial logs, chemical starting fluid, and bricks—they, too, add undesirable flavors. In a pinch, you can start the fire with paper or soft wood shavings, but take care that they have burned thoroughly before adding the hardwood. The best starting fuel is already burned hardwood scraps or hardwood kindling.

Maintaining the Fire

A smoking fire should not be a consuming blaze. Few pieces of meat, fish, or poultry require temperatures of much above 120°F.

A smoking fire should be going strong for 45 to 60 minutes before the food is put into the smokehouse. Before you start the smoking process, observe the fire for a while to see how the wood you are using burns—how much draft will be required, how quickly the wood burns—in short, try to understand the nature of the fire before you begin smoking.

As a general guide, 2 bushels of corncobs or a pile of wood 2 by 4 feet will produce a fire for 4 days of smoking.

"9"

CURING–HOW TO
USE THE DRY METHOD
OR BRINE

CURING–HOW TO USE THE DRY METHOD OR BRINE

Curing is the treatment meat, poultry, and fish are given prior to smoking. Corned beef and salt pork are the only two preserved meats that are eaten without smoking, and several kinds of fish may be, but most people prefer *cured and smoked* food. North and South American Indians have cured meat on the spot as they catch it for hundreds of years. Lacking refrigeration, they soon learned that this was the best way to preserve a catch.

There are two basic kinds of curing: dry and brine. Both accomplish the same purpose: they impart a special flavor to the food, and they help to retard bacterial growth and enzymatic action. Salt, which is the essential ingredient in both brine and dry curing, draws moisture from the cells of the food and also enters the cells through a process of osmosis. It is the salt in the cells, of course, that imparts the flavor that is unique to cured and smoked foods. The amount of salt and the evenness with which it is distributed through the food are a measure of the success of curing. On the other hand, curing is not a complex process. It is done in much the same way today as it was in colonial and frontier days in the United States and Canada. The methods described in this chapter produce excellent results and still leave you some room for experimenting with various flavors in cured foods.

What Kinds of Foods Are Cured?

Beef, lamb, and even mutton are excellent meats for curing. Americans do not normally hold mutton in high regard, but cured and smoked mutton is considered a delicacy in some parts of Europe and is worth experimenting with. Veal is neither cured nor smoked; it does not have the full flavor of beef. Poultry is excellent cured. Almost all varieties of fish can be cured, and this is one of the easiest foods to begin with. Game has not been cured and smoked much in the United States and Canada, although in Europe cured and smoked venison and reindeer are sought-after delicacies. If you want to experiment, use the recipe for brine-cured beef.

Equipment and Materials

Most of the equipment required for curing is already on hand in most homes. The salt and other chemicals used can be special-ordered from a pharmacist, chemical supply house, or purchased from a farm supply store, hardware store, or home center. Following is a list of the basic equipment needed for curing up to 50 pounds of food:

Scale. Preferably one that weighs up to 25 pounds at a time. This is an absolutely essential piece of equipment since meat, salt, and other chemicals must be accurately weighed to insure success and safety in curing foods.

Containers. Five-gallon, wide-mouthed ceramic crocks are ideal, if you can obtain them. They do not absorb the flavors of food cured in them, and they are large enough to hold the food comfortably. Glass, porcelain, or even plastic containers can also be used, but *do not use plastic garbage cans* (these are never suitable for storing foodstuffs) or metal. The containers need not be new, but they should be sterilized. The traditional container, of course, is a clean wooden barrel, the kind farmers have used for years for just this purpose. Such barrels are scarce these days, and although they work quite well for curing, they have no real advantage over a crockery pot, except possibly the mood of authenticity they can lend to the process. And they can be difficult to keep clean and store.

Cutting board or butcher block surface. This is used for trimming and cutting meat.

Sharp knives. They should be large enough and sharp enough to cut meat with ease.

Wooden spoons. Use these to mix the brine or dry cure.

Covers and weights for curing containers. The kind of covers you need depends upon the climate where you live. If the temperature is low and steady, you may only need to cover the food with muslin and place a weight on top of that to hold the food under the brine cure. If there is any chance that the temperature will drop to below 33°F., however, you should use a wooden or other heavy cover to protect the food from freezing. The weight can be any suitable clean material.

Stockings for poultry. Ready-made stockings can be purchased from a poultry or meat supply house and sometimes from an independent butcher who does a lot of his own cutting. You can also make your own from muslin. During smoking, poultry is placed in these stockings to protect it.

Ingredients

The ingredients used for curing are basically the same for all types of cures. Many people prefer a sugar cure—indeed, this seems to be the one most popular with the American palate. There are, however, some exellent cures using salt and other seasoning ingredients.

Salt. Pickling salt is best for all kinds of curing. Never use iodized table salt for curing. The Morton Salt Company produces an excellent pickling salt, and this and other curing ingredients can be purchased from farm supply stores, home centers, and hardware stores. Except for fish, which requires a fine-grained salt, select a medium-coarse one.

Sugar. Table sugar works quite well. Either cane, beet, or corn sugar can be used in curing.

Sodium nitrate and sodium nitrite. These two ingredients, which many experts feel are essential to successful curing, have recently been under close scrutiny from the USDA for their appearance in commercially produced cold cuts and hot dogs. Yet the most up-to-date information from the Department of Agriculture on curing recommends the careful use of these products, and so they are part of these recipes for curing included here. You should realize that sodium nitrate, more commonly known as saltpeter, and sodium nitrite, can have an adverse affect on the body if not used in proper amounts. Used with care, however, they are quite safe and play an important role in maintaining the pinkish-red color that makes cured and smoked meats appetizing to the eye—and to the palate. They also play a lesser role in slowing bacterial growth.

The secret of using sodium nitrate or sodium nitrite is to measure them very carefully and take care not to exceed the amounts suggested in the recipes.

Again, these ingredients can be omitted entirely from any curing process, but the meat will not have the usual appetizing pinkish-red color associated with curing.

Water. This is the last essential ingredient in curing, and it is only used in brine-curing. The water should be pure drinking water. To be entirely safe, boil it before use, but cool it to room temperature before using in the cure.

Other optional ingredients that are used in curing include spices, herbs, and honey. Honey is particularly used with lean meat, where it replaces the sugar as an ingredient. Pepper is commonly used, and other good spices to use are tarragon, marjoram, bay, thyme, fennel, and coriander—you are limited only by your imagination and taste buds.

Dry Curing vs. Sugar Curing

A dry cure, which consists of salt, sodium nitrate and/or sodium nitrite, sugar, and other optional seasonings, is the fastest way to cure and is best for unpredictable climates that are likely to warm up suddenly.

Sugar-cures produce the most flavor, to most people's way of thinking, but it may simply be that sugar-cured foods are most widely available in grocery stores and people have simply grown used to eating them. Many people who have experimented with other seasonings in salt cures discover they like these as well and often better than sugar cures.

A brine cure is essentially a process of pickling the meat; it can be sweet (with sugar) or unsweetened (without sugar or honey).

It is interesting to experiment with several kinds of cures and quite easy to do since the equipment is the same for both. Keep careful records as you cure so you will know what methods and seasonings best suit your taste.

The Importance of Temperature

The ideal temperature for putting up meat to cure is 38°F. Curing meat should be held at 36° to 38°F.—a fact that explains why curing is traditionally done on crisp fall days. If the temperature goes above 50°F., the meat will spoil. If the temperature goes much below 36°F., the meat will not be good when it is smoked. Meat does not cure at a temperature of 34°F. or lower. In addition, food that has been frozen is not suitable for curing. Be sure the temperature of the brine solution is correct before pouring it over the meat. When using a dry cure, the same temperature (38°F.) should be maintained.

Overhauling

Overhauling is a process of repacking the meat and the curing mixture during curing. It insures that all the pieces of meat will come equally in contact with the curing mixture. If a dry cure is used, a fresh mixture is applied to the food. A brine cure is stirred to mix up all the ingredients and reused. The curing mixture and the meat can be repacked into a new container or returned to the old one.

The frequency of overhauling is usually listed along with the curing recipe, but you will soon learn to use your own judgment on this. Meat that is to be cured for 25 to 30 days should be overhauled 2 or 3 times. Cures that last only 10 to 15 days may only need overhauling once.

Preparation for Curing

The food should be fresh and in perfect condition. Meat and fish should be fairly lean, while poultry should be well fattened. Do not cure poultry that has tears in its skin. Meat should be trimmed to remove excess fat and to make a smooth surface. Bones may be left in during curing and smoking.

Curing is a more exact science than either smoking or drying, although it has less guidelines than canning. *The primary precaution is that all food and ingredients must be accurately weighed.* It is foolish to use excessive amounts of sodium nitrate or nitrite. Excessive use of salt will make meat hard and dry, and too little salt will prompt spoilage.

BRINE CURE FOR BEEF, LAMB, AND GAME

2½ gal. water	¼ oz. sodium nitrite
4 lbs. pickling salt	50 lbs. beef, lamb, or game (all
1½ lbs. sugar	one variety—do not mix
1½ oz. sodium nitrate	meats in curing)

Weigh all ingredients accurately. Boil water and cool to 36°F. Mix dry ingredients with water in a large bowl. Place trimmed meat in a crock or other curing container, putting larger, heavier pieces in first. Do not cure food in a container that is more than 3 feet deep.

Gradually pour in the brine. Cover container and add a weight to keep the food completely submerged. Place container in the refrigerator or any other storage place where the temperature of 36° to 38°F. can be maintained.

A general rule of thumb is to cure red meat, including pork and ham, 4 days per pound for a medium cure, slightly less time for a light cure. Overhaul once, somewhere between the fifth and seventh day of curing. Cure bacon 10 days or 1½ to 2 days per pound. Cure fin fish 12 to 24 hours and shellfish 5 to 10 hours. A dry, salty and quite delicious ham is produced by curing for 30 days and overhauling twice.

DRY-SUGAR CURE FOR HAMS AND OTHER PORK CUTS

4 lbs. pickling salt
1½ lbs. sugar
1½ oz. sodium nitrate
¼ oz. sodium nitrite

50 lbs. trimmed, smooth
hams, bacon cuts, and
other pork cuts

Mix curing ingredients well in a large bowl, taking care to distribute sodium nitrate and sodium nitrite evenly. Weigh mixture again and set aside exactly ½ to use later in overhauling. The remaining ½ portion should be used in proportion to the weight of each cut of meat. For example, a ham that weighs 20 pounds should be coated with $^2/_5$ of the curing mixture. Use a sharp knife to cut around the bones and joints. Rub the mixture into the meat with your hands, using a circular motion. Apply a generous amount of the curing mixture to the joints and bone areas.

Be careful of the amount of salt applied to a piece of bacon, which needs only mild curing. Because its surface area is large, there is a tendency to apply a lot of the curing mixture. This is not necessary since bacon is not a thick cut, and the cure does not have to penetrate very deeply to do its work.

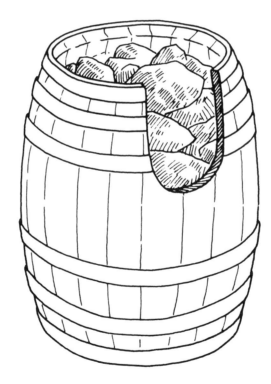

Figure 22. Cutaway barrel, showing meat curing

Dry-cured meat should be packed in a crock or other container that has holes or a rack in the bottom, to allow the liquid that is drawn out during curing to drain away from the meat. To pack food, first sprinkle a small amount of the curing mixture on the bottom of the container. Pack in the meat, putting the larger pieces closer to the bottom. Cover with a cloth and store in a cool place or a refrigerator. Again, the temperature must be maintained around 36°F. After 5 days, overhaul the meat, using the second ½ of the curing mixture. Generally, hams and shoulders should be cured 2 days per pound; bacon should be cured 1½ to 2 days per pound.

Brine Cure for Fish

Any kind of fish can be cured at home, but lean-fleshed white fish work best because the brine—or salt, if you use a dry cure—penetrates their flesh faster than other kinds. Use the freshest possible fish and take care that it has not been frozen. Pickling salt should be used, but choose a fine-grained salt rather than a coarse one.

Split small fish down the back but do not cut through the belly. This will enable them to lie flat, and will also prevent disintegration during curing and smoking. Fish should be thoroughly cleaned to remove all traces of blood. After cleaning, soak them in a mixture of 1 gallon water plus 1 cup salt for 30 minutes to 1 hour. Drain for 10 minutes and pat dry. Score the flesh so the fish will retain a flat shape as it dries.

For larger fish, clean and fillet, removing the backbone, but do not remove the collarbone just below the gills. It serves to support the weight of the fish when they are hung to dry or smoke; without it, the fish will fall apart. Score the flesh side about ¼ to ½ inch deep and 2 inches apart. Remove the skin and fins on thick-skinned fish. This is easily done by placing a sharp knife under the skin and carefully trimming it away. Soak in salted water as described for small fish. Weigh the fish and then weigh out an amount of salt that is approximately ¼ the weight of the fish. (Fat-fleshed fish may require a salt-to-fish ratio of 1:3; you will soon learn to judge the proper amount.) The amount of salt depends upon several factors: less small-grained than coarse salt is used; more salt is used in warm weather, and more salt is used with fat-fleshed or thick fish. On the other hand, too much salt will lower the quality of the fish and make it tough.

Sprinkle a small amount of salt on the bottom of the crock. Place the salt on a flat platter or in a container large enough to hold the fish. Dip each fish into the salt, taking special care to pat the salt into the scored areas. Pick up the fish and any salt that clings to it and pack it into the container skin side down. Pack the fish in even layers, without overlapping any pieces. Packing patterns for small and large fish are shown in Figure 23. Pack the final layer with the skin up. Top the fish with a loose-fitting cover and add heavy weights. The fish makes its own brine as it cures. Small fish cure in 2 days; larger fish require up to 7 days. Remove the fish from the brine, scrub well, and hang to dry before smoking.

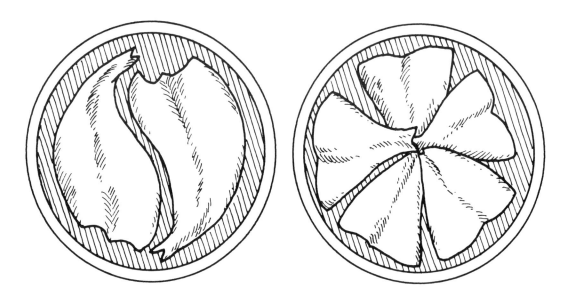

Figure 23. Methods for packing fish in curing pot

Dry Cure for Fish

Trim and clean the fish as described under brine curing. The fish should also be cut and filleted in the same way, taking care not to split the belly in small fish and leaving the collarbone intact in large fish so that the flesh will not fall away during drying. Soak the fish in a saltwater solution of 1 gallon water to 1 cup salt for 30 to 60 minutes. Drain for 10 minutes and pat dry.

The ratio of salt to fish for dry curing is 1:4. Dredge the fish as for brine curing.

The only difference between brine-curing and dry curing is that, in dry curing, the fish are placed so that the liquid can drain away. Figure 24 shows a drying and curing shed that is suitable for this type of curing. Please note that, just as fish must be dried in the shade, they must also be cured in it. A smaller, modified version of this shed can easily be built or drying trays can be used. Arrange the fish flesh side up, alternating the wide and narrow ends. Sprinkle with a light layer of salt. The fish can be stacked if necessary. Place the top layer with the skin side up. Fish should dry cure for 2 to 7 days, depending upon the weather and the type of fish.

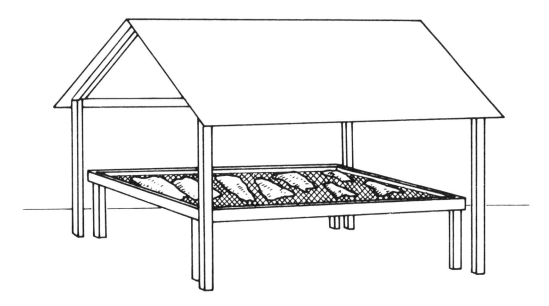

Figure 24. Drying and dry-salting rack for fish

BRINE CURE FOR POULTRY

6 qts. water, boiled and cooled
   to 38°F.
2 lbs. pickling salt
1 lb. white or light-brown
   sugar

1½ oz. sodium nitrate or ⅔
   oz. sodium nitrite
1 bay leaf
¼ oz. fresh black
   peppercorns
25 lbs. poultry

Mix brine ingredients and cool to 36° to 38°F. Pack birds in crock or other suitable container. Pour brine over birds. Cover and weigh down to keep birds submerged. Store in refrigerator or other suitable cool place. Cure for 2 days.

Remove from brine and soak in cold water for 1 hour. Hang to dry. Place poultry in stocking and follow smoking directions for poultry.

PASTRAMI CURE FOR BEEF

4 cups water	2 onions, chopped
15 Tbs. salt	2 garlic cloves
5 tsp. brown sugar	1 tsp. sodium nitrate
¾ tsp. fresh black peppercorns	1½ lbs. beef shoulder, choice
¼ cup chopped fresh parsley	grade

Mix brining ingredients thoroughly. Place beef in a container large enough to hold it and brine. Pour over brine, which should completely cover the meat. Place in refrigerator or in storage area where temperature of 36°–38°F. can be maintained. Cure for 7 days. Rehaul on fourth day, taking care to stir brine before pouring it over the steak again. Smoke pastrami according to directions in Chapter 10.

After the Cure

All cured food should be scrubbed and soaked to remove excessive saltiness. It should be hung to dry thoroughly prior to smoking. Directions for soaking vary with the kind of food and are given in the next chapter, as are directions for smoking the individual kinds of foods.

“10”

HOW TO
SMOKE MEAT, FISH,
AND POULTRY

HOW TO SMOKE MEAT, FISH, AND POULTRY

Although cured food technically may be kept in its brine or dry cure until ready to eat, most people prefer to treat it with the final step—smoking. Some people feel that smoked foods are an acquired taste; whatever they are, their taste is unique and in most cases far more subtle than that produced by regular cooking.

Meat, fish, and poultry may be hot smoked or cold smoked. Hot smoking, which is really barbecuing, is done at temperatures over 100°F. and in light smoke. The drying process at this temperature is faster than during cold smoking. Food that has been hot smoked does not keep very long. It should be eaten right away or within 1 to 2 weeks.

Cold smoking, on the other hand, produces a smokier flavor and keeps for a period ranging from several weeks to several months. A properly smoked ham will even keep for up to 1 year, and Smithfield and Virginia-style cured hams keep up to 2 years.

Food is cold smoked at temperatures of 70° to 100°F. in dense smoke. Cold smoked food is frequently aged for one to three months before eating, although this aging process is not necessary but is only meant to add a subtle flavor and taste.

After curing, food should be scrubbed in warm water. If for any reason the curing has taken longer than expected, the food should be soaked for a few hours to remove excessive saltiness. Fish does not require this extra soaking, but it should be scrubbed to remove salt and dirt. Hang the cured food to dry in an airy, shady place for 24 hours if it is meat, or for 3 to 4 hours if it is poultry or fish. In both cases, a shiny, somewhat tough coating will form on the outside when the food is properly dried. An electric fan is useful is speeding this drying.

When to Smoke

Smoking is a seasonal activity—it works best in cold weather, and many farmers put aside time each fall or early spring for stocking their smokehouses.

A temperature between 35° and 50°F. is ideal. Only experienced smokers should attempt to smoke in zero or subzero weather.

The process of smoking usually requires several days. Ideally, the smoke fire should be maintained day and night; realistically, the quality of the food will not be drastically altered if the drying is done only during the day. And if the temperature should take a sudden drop to zero or lower, by all means halt the smoking, refrigerate the food, and continue drying when the temperature goes up again.

How to Hang Meat and Poultry

Hang all cuts so that the largest surface is exposed to smoke. No cut should be touching any other or any wall of the smokehouse. This means hanging hams and shoulders by the shank end. Using a sharp knife, make a small incision and push a piece of steel wire through it.

Poultry is easier to handle if it is wrapped and tied in a muslin or cheesecloth stocking; the wire is strung under the wings and tied high in back. Poultry is hung breast-side down.

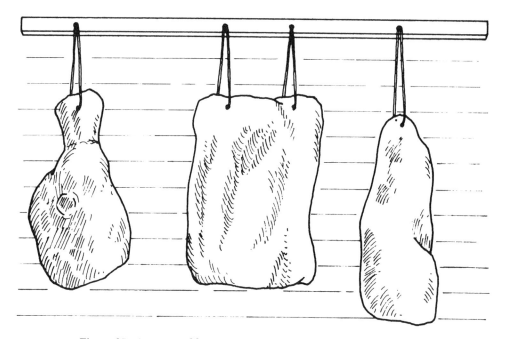

Figure 25. Meat smoothly trimmed and wired, ready for smokehouse

How to Smoke Beef

The best smoking cuts are choice grade. The shoulder is the traditional cut for smoking, although there is no reason why you could not experiment with other cuts if you desire. Smoke meat 2 to 3 days at 100° to 120°F. Properly smoked meat is firm and mahogany-colored. Smoked beef is not usually aged and may be eaten anytime after smoking. It becomes harder as it gets older.

How to Smoke Pastrami

Pastrami is a highly spiced beef shoulder that has been cured in a special solution (see Chapter 9) that imparts a distinct flavor to the meat. Smoke the beef at 90° to 100°F. for 3 hours. Bring the temperature to 180°F., open the smokehouse vents, and smoke the meat lightly for a maximum of 4 more hours. Pastrami may be served at once or stored for future use.

How to Smoke Lamb

Smoke a choice cut of either shoulder or leg. Lamb should be smoked for two days at 100° to 120°F., or until it turns a deep mahogany color. For a smokier flavor, the lamb can be smoked either continuously or intermittently for several more days.

How to Smoke Pork

Smoke a shoulder or loin at 70° to 90°F. for 1 to 4 weeks. Hang in a dark, cool place to age for several months. The same cuts may be hot smoked for about 2 days in a temperature of 100° to 120°F. Open the ventilators of the smokehouse to insure light smoke during this process.

How to Smoke Bacon

Build a very slow, almost smothered fire before putting a slab of bacon in the smokehouse. Smoke at 90° to 100°F. for 5 to 10 days. When hanging bacon, it is sometimes best to put 2 pieces of steel wire through it so it will hang—and smoke—without folding. Wrap in muslin and age a maximum of 3 months.

How to Smoke Sugar-Cured Ham and Pork Shoulder

Smoke at 90° to 100°F. over a slow, almost smothered fire for 2 days. Reduce the temperature to 80°–90°F. and continue smoking for 1 to 2 weeks or longer, until the meat turns a light chestnut color.

How to Smoke Poultry

Chicken is frequently hot smoked, or barbecued, but it is never cold smoked. On the other hand, ducks, geese, turkey, and game birds are excellent cold smoked.

After removing the bird from the cure, drain all cavities and pat the bird dry with paper towels. Dry in an airy, shady place for several hours. Cut large birds into quarters or halves. You may want to wrap the bird in cheesecloth or muslin prior to drying; string the steel wire around the outside of the muslin.

Smoke at 90° to 100°F. in dense smoke for a number of hours that is equal to approximately ½ the bird's weight. To finish, raise the temperature to 180°–200°F. and hot smoke the same number of hours as bird was cold smoked, or until bird turns a dark gold color.

An alternate method is to smoke at 135° to 140°F. for 18 to 20 hours, or until the dark gold color is reached. Age for four weeks at a temperature of 68°F. in a cool, dark place. To serve, cook as you would an unsmoked bird—bake, roast, or combine with other foods.

How to Smoke Fish

Smoked fish is an excellent way to try out the smoking process. If the curing and smoking are properly done, a high-quality product is virtually guaranteed.

Fish, like meat or poultry, can be hot smoked or cold smoked. It keeps in proportion to the time it is smoked—that is, cold-smoked fish may be kept for several weeks under refrigeration and hot-smoked fish lasts only a few days and really tastes its very best when eaten immediately after smoking. Hot smoking works for most varieties of fish, but cold smoking is best reserved for lean fish such as sea herring, alewives, butterfish, and larger species of lean whitefish.

Remove the fish from the cure. Soak and scrub off any dirt or salt. If fish has been cured in brine, pat dry with paper towels. Hang in a *shady*, airy place to dry for about three hours, until a shiny outer skin called pellicle forms.

Fish is smoked at lower temperatures than meat or poultry—rarely at more than 90°F. The vents should be open so the smoke will be light.

In warm, dry climates such as that of southern California, the temperature for smoking should be held at about 90°F.; in more humid northern climates, 70°F. is the best temperature. Maintain either of these temperatures for 24 hours if you want the fish to keep for more than 2 weeks, then close the vents to produce a dense smoke and continue smoking the fish for 5 to 10 days more. For fish that will be eaten within two weeks or less, maintain the temperature for 8 to 12 hours with light smoke. Then close the vents to produce dense smoke and continue to smoke, for a total time of 24 hours.

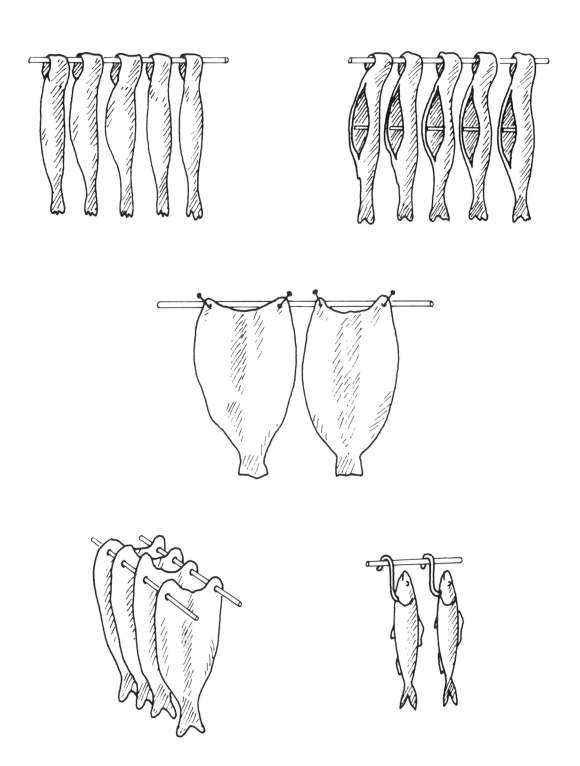

Figure 26. Methods of hanging fish for smoking

Storing Smoked Foods

Smoked meat has two natural enemies: light, which causes rancidness, and insects. The work of both enemies can be halted through careful storage.

Cool meat after smoking. Wrap it in a parchment paper that is thick enough to absorb any extra fat. Then slip it into muslin bags and tie with string at the top. Do not pull out the wire used for hanging around the outside of the muslin bag because insects can crawl down it into the bag. Label each piece of food with the dates of curing and smoking. Although some foods improve considerably with age, most foods are best if eaten within a few months after smoking. Bacon especially does not have much longevity, and it should be eaten within 3 months. Store the meat in an airy, dark place once it has been properly wrapped.

This period of storage is also a period of aging for certain cuts of meat—namely hams and pork shoulder. Up to 1 year, these cuts are improved in flavor and quality through aging. Only Virginia and Smithfield hams, which have received a special cure, benefit from a full 2 years of aging.

Food that is aged for any length of time may develop mold. Light mold is easily removed with a damp cloth, and heavier mold should be cut away. This kind of mold does not affect the quality or edibility of the food.

Smoked fish and poultry should be carefully wrapped and refrigerated. Smoked poultry can also be frozen if you want to keep it for several months.

Recipes That Use Smoked Meat or Fish

SMOKED GOOSE PUDDING

This pudding, which is the equivalent of the traditional Yorkshire pudding, has a slightly more exotic flavor.

bread crumbs	*1½ cups smoked goose,*
4 Tbs. chicken drippings	* ground or finely minced*
4 Tbs. flour	*½ tsp. salt*
⅔ cup water	*¼ tsp. freshly ground pepper*
8 eggs, separated	*¼ tsp. dried dill*

Preheat over to 350°F. Lightly butter a 1½-quart mold and sprinkle bread crumbs in it. Turn upside down and shake to remove excess crumbs. In a small saucepan, combine chicken drippings, flour, and water. Cook over low heat, stirring constantly, until slightly thickened and smooth. Cool. Pour into a large mixing bowl and add egg yolks, smoked goose, and seasonings. Stir to mix well. In a separate bowl, beat egg whites until stiff. Carefully fold into goose mixture. Pour pudding into the mold. Cover with buttered waxed paper. Set pudding mold in a pan of steaming water and bake on middle shelf of oven for 1 hour. Unmold and serve hot. Serves 6.

CHOLENT

This is a dish traditionally served at the luncheon meal that ends the Sabbath for Jews. There are many versions of it, but one of the better ones combines smoked goose and beef.

1 lb. dried white beans	1 lb. smoked beef, cut into
2 Tbs. chicken drippings or	bite-sized pieces
shortening	1 cup smoked goose, cut into
1 large onion, minced	bite-sized pieces
3 garlic cloves, minced	½ tsp. salt
2 Tbs. flour	½ tsp. freshly ground pepper
1½ Tbs. imported paprika	

Soak beans in 1 quart water overnight. Melt drippings in a large pot and sauté onion and garlic in it. Stir in flour and paprika and cook over medium heat for 5 minutes, stirring constantly. Add 2 cups of water from beans and stir until smooth. Add beans, meats, and salt and pepper. Bring to a simmer, then reduce heat to very low. Cover and cook for about 2½ hours, until meat is tender. Add small amounts of water as needed to keep moist. Serves 8.

SOPA DE LIMA

This is a slightly modified version of the delicious Mexican-Mayan soup served on the Yucatan Peninsula. The unusual flavor comes from the addition of smoked pork.

3 cans chicken stock	4-6 Tbs. vegetable oil or other
4 cups water	fat
juice of 1 lemon	6 corn tortillas, cut into ¼-in.
juice of 1 lime with pulp	strips
⅓ cup chopped fresh parsley	1½ cups smoked pork, cut into
⅛ tsp. crushed dried red chili	¼-in. strips
pepper	4 scallions, minced
½ tsp. chili powder	

Combine chicken stock and water in a large pot. Add the lemon juice, lime juice and pulp, parsley, pepper, and chili. Bring to a simmer and cook 10 minutes. In a large skillet, heat oil and sauté the tortillas until crisp. Immediately before serving, add tortillas, pork, and scallions to soup. Heat through and serve at once. Serves 6.

FRIZZLED BEEF

This old traditional American recipe using dried beef still makes a good winter luncheon dish today.

4 Tbs. butter
¼ lb. shredded dried beef
3 Tbs. flour

2 cups milk
½ tsp. salt
¼ tsp. freshly ground pepper

Heat 2 tablespoons butter in a large skillet; add beef and fry over medium heat until it is crisp and curled. In a saucepan, melt remaining butter; stir in flour and cook, stirring constantly, 5 minutes. Add milk and cook, stirring constantly, until white sauce thickens. Add salt and pepper, adjusting amount to taste. Stir in frizzled beef. Serve hot over toast. Serves 4.

SMOKED CHICKEN-BARLEY-MUSHROOM CASSEROLE

¼ cup vegetable oil
2 large onions, chopped
1 cup barley
2 cups chicken stock
1 cup fresh mushrooms
¼ cup finely chopped dried
    mushrooms

½ cup water
1 tsp. dried tarragon
½ tsp. salt
¼ tsp. freshly ground pepper
1 cup smoked chicken or
    duck, cut into bite-sized
    pieces

Heat vegetable oil in a large skillet and cook onions in it until they are limp. Add barley and cook 5 minutes, stirring occasionally. Add the stock, cover, and cook 30 minutes over low heat. Add fresh and dried mushrooms, tarragon, salt, and pepper, along with chopped poultry. Add water, cover, and cook until all water has been absorbed. Serves 6.

SMOKED HAM COOKED WITH DRIED APPLES AND DUMPLINGS

4 cups dried sweet apples
1 10–12-lb. smoked ham
4 Tbs. brown sugar

DUMPLINGS
1½ cups sifted all-purpose
    flour
3 tsp. baking powder
½ tsp. salt
1 Tbs. butter
¼ cup milk
1 beaten egg
4 cups boiling chicken broth

Place dried apples in a bowl and cover with cold water. Soak for 2 hours. Drain and pat dry. Preheat oven to 300°F. Bake ham at this temperature 18-20 minutes per pound or until a meat thermometer inserted in the ham registers 160°F. When ham has 1 hour left to cook, add apples and brown sugar to roasting pan and stir. Coat ham with mixture formed by brown sugar and fat. Continue baking for last hour. Place ham and apples on a platter and keep warm while you make dumplings.

Sift flour, baking powder, and salt together. Cut in butter and add enough milk to make a soft dough. Add egg. Drop dumplings into chicken broth that has been brought to a boil. Cover tightly and simmer 10–12 minutes, until dumplings rise to surface of pot. Arrange dumplings on platter with ham and apples. Serve at once. Serves 12.

SMOKED SHRIMP SZECHUAN STYLE

2 Tbs. vegetable oil	*2 Tbs. soy sauce*
½ lb. smoked shrimp	*1 Tbs. catsup*
½ tsp. garlic, minced	*dash of sugar*
½ tsp. ginger root, minced	*1 tsp. cornstarch*
4 scallions, chopped	*2 Tbs. water*
¼ tsp. dried red chili peppers	

Heat vegetable oil in a skillet until very hot. Sauté shrimp, garlic, ginger, scallions, chili peppers for about 2 minutes, just enough to heat through. Add the soy sauce, catsup, and sugar; mix well. Mix cornstarch and water and add, stirring constantly while cooking until mixture thickens slightly. Serve at once with rice. Serves 4.

SMOKED SHELLFISH OMELET

Smoked fish can also be substituted for the shellfish.

4 eggs, lightly beaten	*¼ tsp. freshly ground pepper*
½ cup smoked shellfish (crab	*1 Tbs. scallions, chopped*
or lobster), flaked	*1 Tbs. fresh parsley*
dash Tabasco sauce	*3 Tbs. butter*
½ tsp. salt	

Combine eggs, shellfish, Tabasco, salt, pepper, scallions, and parsley in a bowl and beat with a fork for a couple of minutes. Melt butter in an omelet pan and pour in egg mixture. Use a fork to stir the bottom of the mixture as the omelet cooks. Several times during cooking, lift omelet to let some of the mixture run underneath it. When top is creamy and almost set, slide omelet onto a heated plate and fold in half. Serves 2.

SMOKED SHRIMP WITH MUSHROOMS AND SCALLIONS

The strong flavor of smoked shrimp plays off well against the delicate flavors of mushrooms and scallions.

1 Tbs. olive oil	¾ cup fresh mushrooms,
1 Tbs. butter	sliced thin
12 scallions, chopped	2 Tbs. dry vermouth
½ lb. smoked shrimp	salt and pepper to taste

Melt olive oil and butter in a large skillet. Sauté scallions for 1 minute. Add shrimp and cook 3 minutes. Add mushrooms and cook 2 minutes, or until barely limp. Add vermouth and seasoning to taste. Stir mixture carefully while cooking for 3 minutes. Serve hot as a first course. Serves 3-4.

DEVILED FISHCAKES (SALTED FISH)

These delectable little fishcakes can be made with any lean white-fleshed salted fish. Cod is particularly delicious.

½ lb. salted fish	½ tsp. salt
1 cup mashed potatoes	¼ tsp. freshly ground pepper
1 egg plus 1 egg yolk	vegetable oil or butter
½ tsp. hot dry mustard	

Soak fish overnight in cold water. If it is still too salty, place in a pan of water, bring the water to a boil, and drain fish. Repeat this procedure until water appears almost fresh after boiling and fish does not taste too salty. Flake fish; place in a large bowl with potatoes, egg, and seasonings. Use a fork to mix well. Shape into 8 fishcakes. Heat oil or butter (or a combination of both, which prevents burning) in a large skillet and sauté fishcakes in it until lightly browned. Serve hot. Serves 4.

SMOKED FISH KULEBIAKA

A Russian kulebiaka is a mixture of fish with rice and seasonings, baked in a pastry shell. Although it is traditionally made with salmon, it is delicious with smoked whitefish.

PASTRY:	FILLING:
1 cup chilled butter	1 cup milk plus 1 cup water
4 cups flour	2 lbs. smoked whitefish
4 Tbs. shortening	1 tsp. salt
½ tsp. salt	½ tsp. freshly ground pepper
10–12 Tbs. iced water	3 tsp. dried dillweed
1 egg yolk	3 Tbs. lemon juice
2 tsp. water	1 cup long-grain rice
	2 cups chicken broth
	10 Tbs. butter
	½ lb. mushrooms, diced
	2 cups onion, diced

Cut all but 2 tablespoons of butter into flour until all pieces of flour are coated. Cut in shortening and add salt. Toss 8 tablespoons water over the mixture and blend quickly with hands; gather into a ball. Add remaining ice water as needed, in small quantities. Dust pastry ball lightly with flour, wrap in waxed paper, and refrigerate 3 hours.

Bring milk and water to a boil. Add fish and cook over very low heat for 10 minutes. Drain fish, pat dry, and cut into small pieces. In a large bowl, combine the fish, salt, pepper, and dried dillweed, and sprinkle lemon juice over it. Cook rice in chicken broth until it has completely absorbed broth. Chill slightly and add to fish mixture. Heat 3 tablespoons of butter and sauté mushrooms in it until barely limp. Add mushrooms and butter to the fish. In the same pot, heat 5 tablespoons butter and cook onion until transparent; add to fish mixture. Toss all ingredients lightly until well mixed; adjust seasoning to taste.

Divide the pastry in half. Roll out pastry on a floured surface to form a rectangle 8 by 16 inches; pastry should be about ⅛ inch thick when rolled out. Carefully transfer pastry to an oiled baking sheet. Mound filling in pastry, leaving about a 1-inch border all the way around. Brush edges of dough with egg yolk that has been beaten with 2 teaspoons water to seal the edges; reserve part of egg mixture. Roll out remaining half of the dough and arrange carefully over the filling. Crimp edges of the bottom pastry over top and press together to seal. Cut two 2-inch circles about equidistant from each other on the top of the pastry. Chill 30 minutes, while you preheat oven to 400°F. Melt remaining 2 tablespoons butter and pour 1 into each top hole. Brush top with the remaining egg yolk. Bake 50-60 minutes, until crust is light brown. Serve hot or cold. Serves 8-10.

WHITEFISH À LA NORMANDE

Normandy is the land of cream and of Calvados, an apple brandy that adds a mellow flavor to so many Norman dishes. This dish substitutes apple cider, which is cheaper and just as delicious.

3 lbs. smoked whitefish	½ lb. mushrooms, sliced
½ tsp. salt	1½ cups heavy cream
⅛ tsp. freshly ground pepper	1 cup fish stock or clam broth
¾ cup cider	4 Tbs. chopped parsley
3 shallots or green onions, minced	

Lightly oil a baking dish just large enough to hold fish. Soak fish several hours in cold water to remove some of the saltiness. Drain and pat dry; sprinkle with salt and pepper. Arrange the fish in an oiled baking dish. Pour cider over fish and arrange shallots and mushrooms around it. Add cream and stock. Place in a 400°F. oven and cook about 10 minutes, until liquid begins to boil. Reduce to 350°F. and cover with foil. Cook 25–30 minutes. Serve with parsley on top. Serves 6.

SCALLOPED SMOKED FISH

Any smoked fish is delicious in a cream-based dish—there seems to be a natural affinity between the two flavors.

2 lbs. smoked whitefish, cut
   into bite-sized pieces
8-10 dried mushrooms,
   chopped
¼ cup onion, diced
2 Tbs. butter
2 Tbs. flour

1½ cups milk
½ tsp. salt
⅛ tsp. freshly ground pepper
½ tsp. dried thyme
⅓ cup grated Parmesan
   cheese

Soak fish for several hours in water to remove some of the saltiness. Drain and pat dry. Soak mushrooms in cold water for 1 hour. Drain and squeeze out excess water. Make a white sauce: melt butter in a saucepan and stir in flour. Cook for 5 minutes, stirring constantly, over medium-high heat. Add milk and stir until sauce thickens enough to coat spoon. Add salt, pepper and thyme. Stir in the drained fish, mushrooms and onions. Spoon into individual serving dishes or one large oven-proof dish. Sprinkle the cheese over the top and bake for 10 minutes at 350°F. Run under broiler for 1 minute to brown. Serves 4-6.

COD PUREE

This dish, unusual to an American palate, is a traditional Lenten dish in southern France.

2 lbs. smoked cod
2 large cloves garlic, minced
2 cups olive oil or olive and
   vegetable oil, mixed
1½ cups light cream
½ tsp. salt
¼ tsp. freshly ground black
   pepper

¼ tsp. nutmeg
1 large dried mushroom,
   soaked 1 hour, squeezed dry
   and diced
4 pieces bread, halved, crusts
   trimmed
4 Tbs. butter

Soak cod overnight, changing water 2 or 3 times. Place in a large pot, cover with boiling water, and cook at a simmer 10 minutes. Drain and pat dry. Flake fish and put into a saucepan. Stir in garlic. Heat the oil and cream in separate pans. Stir about 3 tablespoons of oil and cream together and add to fish. Place fish over medium-high heat and cook, stirring constantly, until fish and oil are combined thoroughly. Stir in remaining oil and cream, alternately, in small amounts, until ingredients blend and puree is formed. Immediately before serving, add seasonings. Sauté toast in butter and serve the puree on it. Serves 4.

SMOKED FISH SOUP

The zesty flavor of a Mediterranean soup is only improved by the addition of smoked fish.

8 cups water
2 lbs. smoked fish
2 Tbs. vegetable oil
4 oz. smoked ham, diced
1 onion, coarsely chopped
1 cup pitted black olives,
  chopped
1 large potato, diced

1 16-oz. can tomatoes
8 Tbs. fresh parsley, chopped
1 tsp. dried thyme
1 dried bay leaf
pinch saffron (optional)
1 tsp. salt
½ tsp. freshly ground pepper
½ cup vermouth

Fill a large soup pot with the water, add fish and simmer 10 minutes. Remove and pat dry; cut fish into large chunks and set aside. Heat vegetable oil in a soup pot and sauté ham in it about 5 minutes. Add remaining ingredients one by one, stirring to mix together. Simmer 30–40 minutes, until potato is cooked through. Serves 8.

MUSHROOMS STUFFED WITH SMOKED MEAT

This appetizer is an excellent way to use leftover smoked beef, lamb, or pork.

12 large mushrooms, clean
  and stemmed
1 cup cooked, smoked meat
1 cup bread crumbs

8 Tbs. butter
2 Tbs. dried marjoram
½ tsp. salt
¼ tsp. freshly ground pepper

Mince mushroom stems and mix with smoked meat, bread crumbs, melted butter, and seasonings. Fill each mushroom cap with mixture. Bake in a 375°F. oven about 30 minutes. Serves 4.

DRIED MUSHROOMS À LA FORESTIÈRE

This makes an unusual brunch dish or a light supper dish.

20 dried mushrooms (cèpes, if
  available)
5 Tbs. butter
6 slices smoked bacon, cut
  into 1-inch cubes

8–10 red potatoes, cooked
  and peeled
½ tsp. salt
¼ tsp. freshly ground pepper
3 Tbs. chopped parsley

Soak mushrooms in cold water for 1 hour. Drain and squeeze gently to remove excess moisture. Sauté the mushrooms in 2 tablespoons butter for 5 minutes. Drain. Melt remaining butter and fry bacon in it until barely brown. Add the mushrooms, cover partially, and cook over low heat for 15 minutes. Cut potatoes into bite-sized pieces and toss in the mushroom-bacon mixture. Cook about 5 minutes more. Add salt and pepper. Serve hot, sprinkled with parsley. Serves 8.

BEEF ROLL STUFFED WITH SMOKED MEAT

1 12-by-6-by-½-in. beef slice,
*   2–2½ lbs.*
4 Tbs. Dijon mustard
½ tsp. dried thyme
¼ tsp. coriander
½ tsp. salt
¼ tsp. freshly ground pepper
6 Tbs. vegetable oil
1 8-oz. pkg. frozen spinach,
*   cooked and drained*
1 large onion, minced

1 egg
1 large garlic clove, minced
4 Tbs. bread crumbs
1 cup smoked ham, diced
½ cup cooked smoked roast
*   beef, ground fine*
1 medium-sized onion, sliced
1 carrot, chopped
1½ cups beef broth
1½ cups dry vermouth

Spread meat slice with mustard and sprinkle with dried thyme, coriander, salt, and pepper. Chop the spinach finely. Sauté minced onion in 2 tablespoons of the vegetable oil. In a large bowl, combine the spinach, onion, egg, garlic, bread crumbs, ham, and smoked beef; stir to mix thoroughly. Spread this mixture over the large meat slice. Roll carefully and tie with string.

In a Dutch oven or other large pan heat remaining vegetable oil; brown onion and carrot slices slightly in it. Gently transfer meat to pan and brown on all sides. Pour in broth and vermouth; cover and cook on top of stove until meat is tender, about 2 hours. Add water during cooking, if necessary. Serves 4.

SMOKED SALMON DIP

1 pkg. cream cheese
6 slices smoked salmon, cut
*   into small pieces*

1 tsp. lemon juice
2-4 Tbs. heavy cream

Bring cream cheese to room temperature. With a fork, beat salmon, lemon juice, and cream into it. Shape into a ball. Chill well. Serve with crackers. Serves 6–8.

SMOKED BEEF POT PIE WITH DRIED HERBS

¼ cup flour	2 cups water
1 tsp. salt	1 bay leaf
½ tsp. freshly ground pepper	¼ tsp. dried thyme
1 lb. smoked beef, cut into bite-sized pieces	1 tsp. dried basil
	2 cups cooked potatoes, peeled and cubed
¼ cup salad oil	
2 Tbs. chopped dried onions, soaked and squeezed dry	¾ cup dried carrots, soaked and squeezed dry
½ cup chopped fresh celery	pastry for a 1-crust pie

Mix flour, salt, and pepper and dredge meat in it. Heat oil and sauté meat in it on all sides. Add onion and celery and sauté 5 more minutes. Stir in remaining seasoned flour. Add water, stirring to mix well; cook until thickened, stirring constantly. Add seasonings and cook covered 1 hour over low heat. Cool to lukewarm and add potatoes and carrots. Remove bay leaf. Turn mixture into a 1½-quart casserole or a deep pie dish and top with rolled-out pastry. Cut a 1-inch hole in the center so steam can escape. Bake at 400°F. about 30 minutes, until top is browned. Serves 4-6.

FLANK STEAK WITH HERB STUFFING

Italian cooks serve flank steak this way, but they usually have the butcher slice the steak lengthwise, making it a *bracciole*. A good Italian butcher will know how to do this.

1 Tbs. oil	2 Tbs. dried oregano
1 tsp. butter	1 Tbs. dried basil
⅓ cup dried, soaked onion, excess moisture squeezed out	¼ tsp. dried thyme
	½ tsp. salt
	¼ tsp. freshly ground black pepper
1 garlic clove, minced	
⅓ cup pistachio nuts	1 2-lb. flank steak, split or whole, scored on each side
1½ cups bread crumbs	
	1 7-oz. can tomato sauce

Heat oil and butter together in a small pan and cook onions and garlic in it until limp. Pour into a small bowl. Add nuts, bread crumbs, and seasonings. Mix well. Spread mixture in center of the steak (or steaks, if they are split); roll and tie with string. Place in an oven-proof dish. Pour tomato sauce around them. Cook 2 hours, adding water as needed. Serves 4.

EMPANADAS

These are a South American version of meat turnovers.

PASTRY:

3 3-oz. pkgs. cream cheese   1 ½ cups flour
½ cup butter

FILLING:

1 Tbs. salad oil   *dash chili powder*
1 Tbs. butter   *½ tsp. oregano*
½ onion, minced, soaked   *3 Tbs. raisins*
    overnight in water and   *2 Tbs. dried apricots, minced*
    squeezed dry   *fine*
½ lb. ground beef   *2 hard-cooked eggs, minced*
1 tsp. dried basil

Bring cream cheese and butter to room temperature and mix together well. Blend in flour and roll into a ball. Refrigerate about 3 hours.

Make filling: heat the oil and butter in a large skillet; combine all ingredients except eggs and cook 20 minutes, adding slight amounts of water, if necessary. Cool slightly and mix in eggs.

Roll out pastry and cut into circles 6 inches in diameter. Place about ⅓ cup filling on each circle, fold over, and seal the edges. Prick top once with a fork so steam can escape as empanadas cook. Bake on unoiled cookie sheet at 400°F. until lightly browned, about 15 minutes. Serves 4.

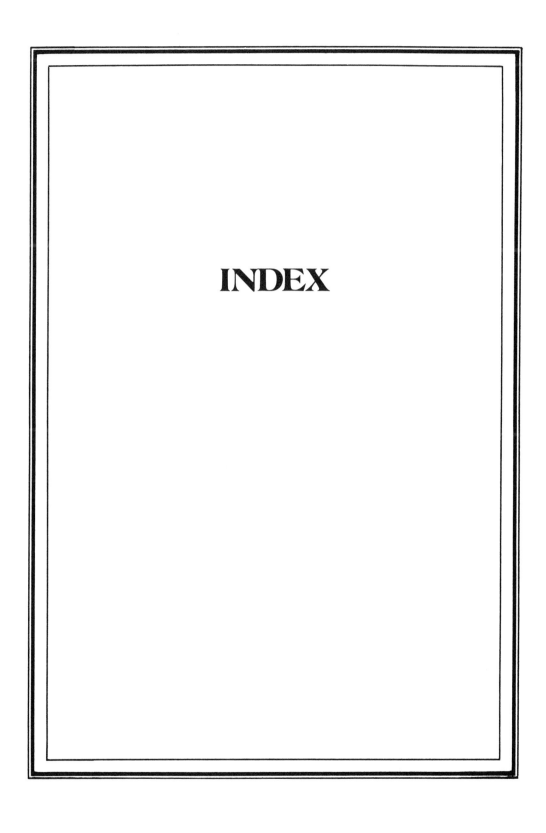

INDEX

INDEX

Illustrations indicated by italics.